Development of Hoeing in Narrow
Seeded Cereals with a Camera Row Guidance

Development of Hoeing in Narrow Seeded Cereals with a Camera Row Guidance

Dissertation to obtain the doctoral degree of Agricultural Sciences (Dr. sc. agr.)

**Faculty of Agricultural Sciences
University of Hohenheim**

Intitute of Phytomedicine,
Department of Weed Science

Spanish National Research Council, Center for Automation and Robotics (CSIC CAR)

submitted by
Benjamin Leon Kollenda

from *Filderstadt*

2019

Bibliografische Information der Deutschen Nationalbibliothek

Die Deutsche Nationalbibliothek verzeichnet diese Publikation in der
Deutschen Nationalbibliografie; detaillierte bibliographische Daten sind im Internet
über http://dnb.d-nb.de abrufbar.

1. Aufl. - Göttingen: Cuvillier, 2020
    Zugl.: Hohenheim, Univ., Diss., 2019

    D100

    ISBN 978-3-7369-7169-1
    eISBN 978-3-7369-6169-2

Summary

The societal expectation to reduce pesticide usage has lately triggered specific research in agricultural science. Especially for weed management in rather narrowly planted cereal crops, there is a need for efficient alternative strategies to reduce the dependency on chemical-based measures. Herbicide applications are still the cheapest, the most effective and the least labour intensive way to control a wide range of weeds and to save the yield potential of cereal crops. This is true, even though numerous resistant weed biotypes have occurred, and new herbicidal mode of actions have not been introduced for some time. Thus, a future perspective for weed management based on herbicides is endangered. Mechanical weed control strategies are the only reasonable direct measure to replace herbicides. But the dependency on favourable weather along with suitable soil conditions leads to a more limited time frame for weed control measures. Furthermore, the decreased yield potential, due to wider row spacing, potential risk of crop damages, along with lower driving speeds has so far limited a wide introduction of mechanical weed control as a future alternative in cereal cropping.

This study aimed to combine a hoe and camera-based row guidance in order to optimise hoeing for introducing it into conventional cereal seeding distances down to 125 mm. The camera guided hoe system enhanced the field efficiency by higher operating speeds and thus making it competitive to spraying equipment. Simultaneously, a weed control efficacy on a comparable level to herbicide application should be achieved. Therefore, the tools need to be guided very close to the crop row to decrease the untreated field area as far as possible. Targeted operating speeds with the proposed technique were between 6 to 8 km h^{-1}. A type of tool suitable for high driving speeds was needed. Additionally, it had to be shaped in a way to bury small weeds occurring in the crop row by moving soil into the intra-row space. First, tools from different fields of application were modified for the use in narrow row distances and were tested in pre-trials on handling performance and weed control efficacy. After that, in a two-year trial series, in row distances down to 125 mm, a selection of three tools from the pre-tests were used to select the most effective and versatile hoe shape for the specific task. The weed control efficacy was separately investigated for the inter-row and the intra-row space. To determine the influence of inter-row weeding to the crops a measurement method based on digital image analysis was used.

After the treatment, it recorded the coverage of crop plants with soil compared to manual weed pulling or a herbicide treatment control plot. An A-shaped hoe knife showed the best relation between weed control efficacy and selectivity. It is flat angled and called no-till sweeps. The no-till sweeps can be used with a speed up to 8 km h^{-1}, without moving too much soil into the crop row, even under compacted soil conditions. The trial series showed that a crop soil cover of 25 % needs to be achieved for successful in-row weed suppression. In the next step, two camera systems and a hoe frame with a hydraulic steering unit were adjusted and developed to precisely guide hoe tools in narrow rows. The Tillett and Hague Inter-Row Vision Control (Tillet and Hague Ltd, Silsoe, UK) in combination with the hydraulic pivot arm parallel steering system K.U.L.T. Argus (Kress Umweltschonende Landtechnik, Vaihingen-Enz, Germany) was the most successfully tested combination. A 3 m prototype was built and tested at three different speeds in 150 mm row distance winter-wheat. The test resulted in a maximum inter-row weed control efficacy of 94 % per run. To further enlarge the field efficiency, a 6 m segmented hoe was built with two separated row guidance units for each side, based on the technical know-how of the 3 m version. With this prototype, 1.45 ha h^{-1} can be hoed, even if seeding was performed with a 3 m width seeding machine. In a field trial in 150 mm seeded summer barley, treated with the segmented hoe, weed control efficacy was at the same level as the 3 m version. In order to determine the accuracy of the prototypes, the 3 m hoe, as well as the 6 m segmented hoe, was tested by a tacheometer record during a typical weeding job. The result showed that the prototypes worked within a standard deviation of maximum 22 mm. Additionally, weed control by hoeing, harrowing and herbicide application was combined and compared in a two-year trial series. It could be shown that hoeing in combination with one herbicide leads to almost complete control success.

For mechanical weed control, unsuitable conditions will always be a challenge, as they narrow the time frame for successful weeding without the risk of crop damage. Nevertheless, the implementation of hoeing as part of an integrated weed management approach will become important when herbicide usage is further reduced by legal restrictions. With precision agriculture technology, like the camera row guidance of this study, farmers will be able to use mechanical weed control with higher field efficiency even in narrow seeded crops with the option to reduce herbicide usage.

Zusammenfassung

Die gesellschaftliche Erwartung, den Einsatz von Pestiziden in der Landwirtschaft zu reduzieren, ist aktuell ein wichtiger Auftrag an die agrarwissenschaftliche Forschung. Alternativen zur chemischen Bekämpfung von Unkräutern und Ungräsern in engeren Reihenabständen wie beispielsweise in Getreide müssen jedoch erst entwickelt werden. Herbizidapplikationen sind nach wie vor die kostengünstigste, effektivste und auch am wenigsten arbeitsaufwendige Methode, um Unkräuter in Getreide erfolgreich zu bekämpfen und Ertragspotenziale zu sichern. Das ist so, obwohl der chemische Pflanzenschutz vor großen Herausforderungen steht. Vermehrtes Auftreten von resistenten Unkrautbiotypen und die Tatsache, dass keine neuen Wirkstoffe zugelassen wurden, kann zu Wirklücken beim chemischen Pflanzenschutz führen. Somit ist die Unkrautbekämpfung mit Herbiziden keine zuverlässige Lösung für die Zukunft. Mechanische Unkrautbekämpfungsverfahren sind heutzutage die einzigen direkten Alternativmaßnahmen zu Herbiziden. Da mechanische Verfahren nur unter bestimmten Witterungs- und Bodenbedingungen erfolgreich sind, ist das Zeitfenster für zur Unkrautbekämpfung kleiner. Gleichzeitig schwankt der Bekämpfungserfolg und kann deutlich geringer sein als bei einer Herbizidapplikation. Die bisherige Technik setzt größere Reihenweiten voraus und führt zu einem niedrigeren Ertragspotenzial. Außerdem besteht eine höhere Gefahr, die Kulturpflanze durch eine mechanische Bekämpfungsmaßnahme zu schädigen weshalb die mögliche Arbeitsgeschwindigkeit von Hacken geringer ist. Diese Faktoren verhinderten bisher eine Verbreitung des mechanischen Pflanzenschutzes in der konventionellen Landwirtschaft.

Ziel dieser Studie war es, durch die Kombination eines Hackgeräts mit einer Kamerareihenführung die Genauigkeit der Unkrautkontrolle so zu verbessern, dass der Einsatz einer Hacke auch in Getreide in engen Reihenabständen von nur 125 mm möglich ist. Die automatische Reihenführung ermöglicht schnellere Arbeitsgeschwindigkeiten und damit höhere Flächenleistungen vergleichbar mit einer Feldspritze. Gleichzeitig soll durch eine exakte Werkzeugführung eine verbesserte Unkrautbekämpfung erzielt werden. Dafür sollen die Werkzeuge näher an der Getreidereihe arbeiten und durch Häufeln von Erde in die Reihe kleinere Unkräuter durch Verschütten am Wachsen hindern. Als Ziel wurden Geschwindigkeiten zwischen 6 und 8 km h^{-1} mit der neuen Technik anvisiert.

Im ersten Schritt wurden Hackwerkzeuge aus verschiedenen Anbaubereichen ausgewählt, angepasst und in einem Vorversuch auf Unkrautkontrolleffizienz und Handhabung in engeren Reihen getestet. Anschließend wurden über einen Zeitraum von zwei Jahren drei vielversprechende Werkzeuge getestet, um die Werkzeugform zu finden, die am vielseitigsten einsetzbar ist und die besten

Bekämpfungserfolge in Reihenabständen bis 125 mm ermöglicht. Der Bekämpfungserfolg wurde für den Zwischenreihenraum und den Bereich in der Reihe separat erfasst. Um den Bekämpfungserfolg in der Reihe in Abhängigkeit vom Häufeln zu untersuchen, wurde eine Messmethode basierend auf digitaler Bildanalyse verwendet. Diese errechnet die Verschüttung der Kulturpflanze mit Boden im Vergleich zu einer permanent händisch gejäteten Kontrollparzelle, in der die Getreidepflanzen unberührt bleiben.

Die besten Ergebnisse bei der Unkrautkontrolle und der Selektivität erzielte ein beidseitig verkleinertes, A-förmiges flaches Schar (Flachhackschar). Dieser Werkzeugtyp war einfach einzustellen und für Geschwindigkeiten bis 8 km h^{-1} unter verschiedenen Bodenbedingungen geeignet. Die zweijährige Versuchsserie zeigte außerdem, dass die besten Bekämpfungserfolge in der Reihe bei einer Verschüttung von 25 % der Getreideblattfläche erzielt werden konnten.

Ein weiterer Schritt war der Test zweier kamerabasierter Reihenführungssysteme. Beide Systeme wurden auf die Möglichkeit der Verwendung in engen Getreidereihenabständen untersucht. Die Kamerasteuerung Inter-Row Vision Control (Tillet and Hague Ltd, Silsoe, GB) in Kombination mit einem hydraulisch gesteuerten Parallelverschieberahmen K.U.L.T. Argus (Kress Umweltschonende Landtechnik, Vaihingen Enz, Deutschland) stellte sich als die beste Lösung heraus. Ein Prototyp mit 3 m Arbeitsbreite wurde gebaut und erfolgreich in 150 mm Saatreihenweite getestet. Die Flächenleistung lag bei 1,45 ha pro Stunde und die durchschnittliche Unkrautbekämpfung zwischen der Reihe bei 94 % pro Durchgang. Um die Flächenleistung weiter zu verbessern, wurde eine 6 m Doppelhacke entwickelt mit zwei unabhängigen 3 m Reihenführungssystemen für je eine Seite. Die gleiche Reihenführungstechnik wie in der 3 m Hacke kam zum Einsatz. Auch dieser Prototyp zeigte gleich hohe Unkrautbekäpfungserfolge bei verdoppelter Flächenleistung. Durch eine Aufzeichnung der Steuerbewegung während einer 100 m Arbeitsstrecke unter normalen Einsatzbedingungen wurde mithilfe eines automatischen Tachymeters die Steuerungsgenauigkeit überprüft. Beide Prototypen erreichten eine Abweichung von maximal 22 mm. Des Weiteren wurden in einer zweijährigen Versuchsserie die Unkrautbekämpfung durch eine Kombination von Hacken, Striegeln und Herbizidapplikation verglichen. Dabei konnte gezeigt werden, dass Hacken und die Applikation eines einzigen Herbizides, zu einem nahezu vollständigen Bekämpfungserfolg führte.

Ungeeignete Bedingungen werden immer eine Herausforderung für den mechanischen Pflanzenschutz bleiben, wenn dadurch die Zeitfenster zum erfolgreichen Bekämpfen kleiner sind als bei einer Herbizidapplikation. Trotzdem kann die vorgestellte neue Hacktechnik ein wichtiger Pfeiler des integrierten Pflanzenschutzes werden, insbesondere wenn der Gesetzgeber in Zukunft vorschreibt, den Herbizideinsatz zu reduzieren. Die Weiterentwicklung dieser Studie ermöglicht es auch in engen Reihenabständen auf konventionellen Betrieben, Hacken in Getreide einzusetzen mit der Option Herbizide einzusparen.

Contents

List of Figures

List of Tables

Chapter 1

Introduction

1.1 General Introduction

Since the beginning of agricultural production, pests and diseases have negatively impacted cropping. The presence of unwanted plant species (weeds) inside a crop creates competition, often leading to substantial yield losses (Walker, 1983). Only an effective removal of weeds in the right time window can secure high yield potentials (Welsh *et al.*, 1999). Furthermore, weeds can impede yield and harvest operations. In wheat, the most grown food crop worldwide, weeds can cause up to 18 % - 50 % loss of yield depending on the weed species (Wilson and Wright, 1990; Dierauer and Stöppler-Zimmer, H, 1994; Lemerle *et al.*, 1996; Oerke, 2006). Weeds are also known to be a potential host for crop diseases like insects (Panizzi, 1997), fungal pathogens (Boland and Hall, 1994) and viruses (Christian, 1993), which can negatively affect the yield quality (Heitefuss *et al.*, 2000).

The increasingly rapid growth of the world population, that will need to be fed, is a permanent challenge (Murchie *et al.*, 2009; Calicioglu *et al.*, 2019). Just stabilising current yield levels worldwide will not close the gap between the decline in the agricultural surface and the increasing demand for food production (Martindale and Trewavas, 2008). Proper weed management strategies are one of the key factors to address this problem. While in 1900, in developed countries, around 60 % of the working population was employed in agriculture (Grigg, 1975), today in Europe this workforce declined to just 4 % (International Labour Organization, 2018). In consequence, for any progress in weed management, reduced labour input and enhanced field efficiency, needs to be considered at the same time.

1.2   Weed control methods

Reduction of weed infestations can be achieved with numerous indirect and direct measures. An overview of these strategies is given by Hamill *et al.* (2004). He grouped them in preventive methods, cultural methods, mechanical methods, as well as chemical and biological methods. To reduce the weed infestation, the use of certified weed-seed free crop seeds and clean equipment are the most important preventive method to impede weed seed spread (Walker, 1995). Further on, row spacing and selection of competitive cultivars, as well as an adjusted seed density are methods to reduce the harmful impacts of weeds (Korres and Froud-Williams, 2002; Hamill *et al.*, 2004). A reduction of weed populations can be achieved by diversified crop rotations, with a continues shifting between winter and spring crops, as well as between grass-crops and broad-leaf crops (Young *et al.*, 1994). Such methods can disturb the life cycle of problematic crop-associated weeds by longer periods with bare soil which increase options for mechanical weed control in spring and fall. Furthermore, dense crops prevent certain weeds to emerge and different crops in wider rotations allow the use of diverse herbicidal modes of action (Monaco *et al.*, 2002; Walker, 1995). For such strategies, planting and harvest time needs to be changed (Zoschke and Quadranti, 2002) to prevent major weed emergence within the crop growth period (Anderson, 1994) and to create opportunities to control them at peak emergence. Linked to that are delayed sowing times for certain crops, which may impact the profitability of farms due to a reduced crop yield potential (Monaco *et al.*, 2002). Direct methods against weed infestation in crops are divers too. All sorts of tillage before planting can control established weed populations (Zoschke and Quadranti, 2002). Especially ploughing is an effective measure by burying weed seeds to reduce weed infestation in the following crops or by moving rhizomes of perennial to the top and drying them (Monaco *et al.*, 2002). Whereas other authors report that reduced tillage can lead to an increased occurrence of weed species (Moyer *et al.*, 1994; Tørresen *et al.*, 2003). Some studies showed that tillage during the night can reduce germination stimulation of light-sensitive weed seeds (Melander *et al.*, 2005; Juroszek and Gerhards, 2004). The preparation of a stale seedbed is a method to reduce the soil seed bank of weeds by stimulate weed germination by seedbed preparation and control them afterwards (Rasmussen, 2004).

Post-emergence weed control inside the crop requires methods to secure crop yield potential. Even though it is economically unattractive, the most successful weed control in crops can be achieved by hand pulling of weeds (Monaco *et al.*, 2002; Zimdahl, 2007). Approaches to control weeds less labour intensive are burning weeds with flame machines, destroying them with hot steam (Dierauer and Stöppler-Zimmer, H, 1994; Monaco *et al.*, 2002), kill weeds by flooding (Monaco *et al.*, 2002), with high voltage (Diprose and Benson, 1984), or with laser beams (Mathiassen *et al.*, 2006) or a cover with plastic or organic mulch (Monaco *et al.*, 2002). Biological approaches use living organisms against weed infestation. It is reported as not effective alone and therefore only applied in combination with other methods (Zimdahl, 2007). Today, the most common weed control methods in developed countries are herbicide based or mechanical measures that will be further described in the following sections.

1.2.1 Development of weed management

Historically, hand pulling of weeds or manual weed removal with simple tools has been the only means of controlling weed infestation in crops for centuries. The invention of early forms of the plough was important to reduce weed infestation by turning the soil, and thus, burying weed seeds. At the end of the 19th century inorganic chemicals such as Iron-II-sulphur or copper salts marked the begin of chemical weed control (Wegler, 2013; Hamill *et al.*, 2004). Later, the phytotoxic effect of sodium chlorate was discovered and used as a non-selective herbicide (Wegler, 2013). In the 1930s indole-3-acetic acid as well as 2-naphthoxyacetic were discovered during research on plant growth regulators (Monaco *et al.*, 2002). In the 1940s, synthetic herbicides including artificial plant hormones such as 2,4-dichlorophenoxy-acetic acid (2,4-D) (Van Overbeek, 1947) and 2-methyl-4-chlorophenoxyacetic acid (MCPA) were triggers for a wider use of chemical plant protection compounds (Monaco *et al.*, 2002; Oerke, 2006). From this point onwards, the development and usage of modern selective herbicides evolved quickly and became one of the most important methods to reduce weed infestation and to enhance the productivity in crop production (Bastiaans *et al.*, 2008). In 1970 the herbicidal effect of glyphosate was discovered and turned later on to be the most important herbicide worldwide (Duke and Powles, 2008). Today, there are 116 herbicidal active substances approved in the European Union. Half of them will expire at the end of 2020 (European Commission, 2019). Currently, the conventional crop production, which heavily relies on chemical plant protection, is at risk if no new methods are adopted or new herbicidal mode of actions are discovered (Coble and Schroeder, 2016) and successfully registered. In the last decades, there was no introduction of new modes of action and this is unlikely to change soon (Duke, 2012). Nevertheless, the success of herbicide-based weed management is caused by low labour intensity and a high weed control potential which can be achieved in a wide time window under various soil conditions.

1.2.2 Undesired impacts of herbicides and public concerns

Beside many advantages, chemical weeding has also some drawbacks. Even selective herbicides can induce stress to the crop and thus may reduce crop yield and yield quality (Salzman and Renner, 1992). This might be acceptable, as long as the damage is smaller than the yield loss due to weed competition. But there are further critical aspects of herbicides. Soon after their introduction, concerns were expressed about the possible selection of herbicide-resistant weed biotypes. Compared to fungicides and insecticides, the use of herbicides seemed to be less affected by resistance development in weed populations, due to the reduced vitality of resistant weed biotypes (Gressel and Segel, 1978). However, the first resistance case of *Daucus carota* L. (wild carrots) to 2,4-D was reported by Switzer (1957). Nevertheless, most of the resistance issues arose from ineffective usage of herbicides and ignoring best management practices such as herbicide rotation (Massa *et al.*, 2013). The continuous use of the same herbicidal mode of action until it is no longer effective, lead to a rapid increase of resistance cases from 1975 onwards (Chauvel *et al.*, 2012; Shaw, 2016). According to current studies by (Heap, 2019), 23 out of 26 known mode of action have been compromised due to the selection of herbicide-resistant weed biotypes. Even though herbicide application can be a very effective weed control method, there are also ecological costs (Slaughter *et al.*, 2008). E.g. the use of herbicides also influences the species composition of the flora on arable land. Studies by Frieben (1990) have shown the reduced occurrence of endangered species, especially in conventional farming areas, while Rydberg and Milberg (2000) reported in a weed survey in Sweden that endangered or rare species were recorded on organic certified farms. Besides that, undesired contamination of non-target areas with herbicides (Nitschke and Schüssler, 1998; Pietsch *et al.*, 1995) may cause further transportation into the environment. Surface water runoff containing herbicide traces after heavy rainfall, spray drift during the application (Squillace and Thurman, 1992; Monaco *et al.*, 2002; Neumann *et al.*, 2002), volatilization after application (Millet *et al.*, 2016) or accidental losses due to leakage of spraying equipment, wrong handling of products, and mistakes during cleaning and disposal (Neumann *et al.*, 2002) are some of the possible sources. Even metabolites of herbicides can be present for a longer time in the environment (Kolpin *et al.*, 1998). Costs of purifying and monitoring of drinking water, to comply with residue limits of pesticides, was estimated by Waibel *et al.* (1998) as not less than 128 Mio. DM per year in Germany. According to reports from the German Federal Ministry for Food and Agriculture (Bundesministerium für Ernährung und Landwirtschaft (BMEL), 2017) there are often residues beyond thresholds, even though the risk of exceeding the maximum daily uptake was not reached in any case. Increasing concerns about food safety and stricter rules about herbicide residues in the food chain, impacted negatively consumer acceptance of chemical plant protection (Hoban, 1998; Wise and Whalon, 2009). Alarmist reports in mass media on the hazards of pesticides, without highlighting the benefits (Cooper and Dobson, 2007), enhanced the demand to renounce chemical plant protection. Although, herbicides still play a dominant role in weed management strategies, there is a strong need for alternative methods. The discussion about herbicides and the fact that chemical weed control reached certain limits, concerning resistance management, generated research programs to reduce the dependency on herbicides. Further developing of mechanical weeding is playing a major role in this respect (Melander *et al.*, 2005).

1.2.3 Mechanical weeding

When used in conventional farming, mechanical weeding is mostly combined with herbicide applications. Also, combinations of a mechanical method and a band-spray application is part of agronomic practices in some crops. Since the use of herbicides is not permitted in organic farming systems, mechanical weeding is the main direct control option beside preventive and cultural measures (Rasmussen, 2004). Mechanical methods achieve weeding mostly by uprooting, cutting or covering weeds with soil (Terpstra and Kouwenhoven, 1981). Pullen and Cowell (1997) separated mechanical weeding systems into two main categories: hoeing and harrowing. Harrowing can reduce weed infestation also in the row between the crop plants (Lötjönen and Mikkola, 2000). But successful harrowing of cereals requires treatments at early crop growth stages (Rasmussen and Svenningsen, 1995). One option is to harrow the first time before crop emergence when weed seedlings just germinated. Brandsæter *et al.* (2012) showed that pre-emergence harrowing lead to a weed density reduction of 26 % and combining of pre- and post-emergence harrowing lead to a weed control success of up to 61 %. If the treatment is done too late only 47 % weed control efficacy was demonstrated (Brandsæter *et al.*, 2012). Inter-row hoeing tools are more effective in controlling especially tap-rooted weeds (Melander *et al.*, 2003). Before the availability of chemical plant protection, hoe-crops such as sugar beets and potatoes were placed in crop rotations in a way to have the possibility to use more aggressive weeding tools like a hoe (Merfield, 2019). Thus, the timing of hoeing treatments is more flexible because weeds can be controlled at later, already well-rooted growth stages (Melander *et al.*, 2003; Tillett, 2005). Another advantage of hoeing compared to harrowing is, that the crop is less mechanically impacted by the tools (Hatcher and Melander, 2003). But depending on the shape of the tools, hoeing can also indirectly reduce intrarow weed infestation by moving soil from the inter-row space into the row, thus covering emerging weeds (Melander *et al.*, 2003). Burying of intrarow weeds works best if the cultivation is performed shallow, in combination with dry soil conditions.

The term hoeing originally was not linked to weeding but was used before the plough was introduced, to describe all type of work to loosen the soil surface. This was performed by hand with simple tools like sticks or handles with right-angled tines at the end. In 1910 E. Hahn (1910) proposed the term of hoe-farming to describe agriculture without ploughing. The invention of the seed drill by Jethro Tull in the 18th century lead to the possibility of inter-row hoeing (Chambers, 1879) on a wider scale. Before that, time weeds were ignored as a potential negative yield impact as they were suppressed by the first mechanical inter-row tillage equipment. Therefore, hoeing is one of the oldest and most common mechanical methods to control weeds in the inter-row space (Griepentrog *et al.*, 2006). But 80 years ago, limitations of direct mechanical approaches lead to a replacement by chemical weeding in developed countries.

1.2.4 Impacts and challenges of mechanical weeding

The major drawbacks of mechanical weeding are various and need to be considered for their improvement. Compared to chemical plant protection, the main challenge of mechanical weed control is the dependence on weather and soil conditions for successful treatments (Kurstjens and Kropff, 2001). Only dry weather conditions in the days after the treatment can prevent uprooted weeds to regrow (Van der Van der Schans *et al.* (2006)). Further on, compared to herbicide applications the treatment costs are higher due to more fuel consumption for the power needed to pull tools through the soil. Additionally, inter-row weed hoeing in cereals requires usually wider crop row distances that lead to reduced yield potential (Holliday, 1963). Selectivity, which describes the ratio of weed control efficacy and crop damage, in mechanical weeding is reached by a growth stage difference between crop and weeds (Johnson, 2002). Concerning hoeing, this is reached by the alignment of the tools to the inter-row space and the adjustment of the tools to the soil condition. Only proper guidance of the tools and an adopted aggressiveness enable a highly selective weed control measure. The possible damages to the crop due to hoeing, requires advanced driving skills which may impact other important crop operations if such skills are urgently needed in the same time window (Jabran and Chauhan, 2018). Hand steering a hoe without damaging the crop is difficult to achieve with working speeds of more than 5 km h^{-1} over a longer time period. Also, the weed control efficacy is limited by the untreated area along the crop rows, due to the need for a safety distance between crop and hoe tool. The weeding effect on perennial weed species is not efficient (Lötjönen and Mikkola, 2000) as they can be just reduced but not sustainably eliminated. According to Melander *et al.* (2012) and Brandsæter *et al.* (2017), a sustainable reduction can be only achieved by a combination of more than one measure, where also direct control methods need to be included. That is why high time consumption and reduced levels of weed control limited the acceptance of mechanical weeding as a standalone tool in conventional farming in general (Pullen and Cowell, 1997). Tillett *et al.* (2002) showed that hand steered hoeing compared to spraying is slow. In contrast to chemical weed control, mechanical weeding always creates a physical disturbance on the topsoil layer. In consequence, mechanically weeded fields are prone to erosion, particularly in hilly landscapes and in the presence of light soils (Van Oost *et al.*, 2006). Specifically, inter-row hoeing causes soil loosening and enhances the danger of erosion in case of heavy rain events. When re-compaction behind the hoe tool is applied, uprooted weeds get in close contact to the soil again, rendering the complete treatment useless in terms of weed control. As hoeing mostly needs numerous passes to achieve full weed control, soil compaction is increased and can cause yield reduction in rows close to the driving tracks (Kouwenhoven, 1997). In case hoeing is conducted early in spring, there is a risk for frost damage of the crop (Van der Van der Weide and Bouma (1997)). Other restrictions of inter-row hoeing compared to herbicide spraying is the strong linkage between seeding width, determined by the sowing equipment and the working width of the hoe.

The row width between neighbouring sowing strips varies compared to the row width within a strip (Figure 1.1).

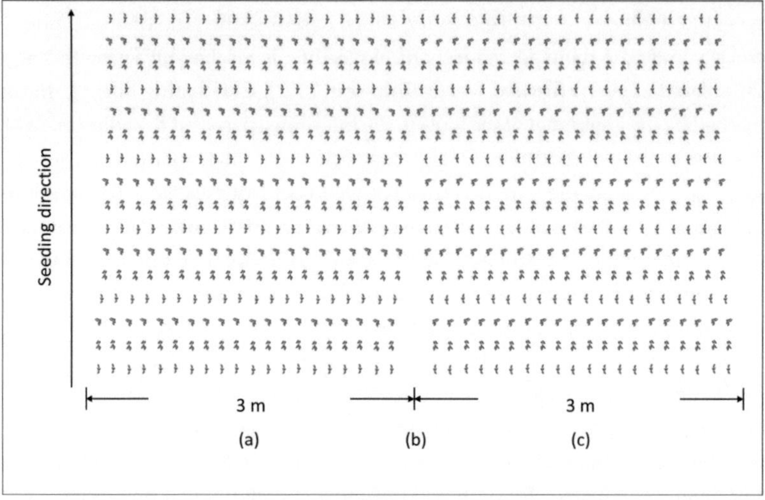

FIGURE 1.1: Typical sowing strip alignment. **(a)** Initial sowing strip, **(b)** Row width deviation of adjacent sowing strips, **(c)** Following sowing strip. (B.Kollenda)

Thus, it is not possible to enlarge the working width of the hoe over more than one seeding strip, due to the spatial deviation of adjacent strips, even if seeding would be performed with Real-time Kinematic (RTK) supported global navigation satellite system (GNSS) guidance. Mechanical weeding and combinations of new possibilities, provided by precision farming technologies, can be a future strategy to address the mentioned issues and to reintroduce it as a major direct measure against weed infestation.

1.3   State of knowledge

Numerous patents were submitted in the first decades of the 20th century describing inventions to mechanically control weeds. These cover simple knives, that were attached to tillage machines (Orr and Radley, 1913), as well as devices for inter-row hoeing close to current systems (Graf Littichau and Liegnitz, 1941). During the mechanisation era, when the tractor was introduced, tillage tools became precise enough to use them also in crop rows (Jabran and Chauhan, 2018). From the 1960s onwards mechanical weed control devices were further developed mainly in support of organic agriculture (Jabran and Chauhan, 2018; Van Der Weide *et al.*, 2008). In cereals, a lot of studies deal with harrowing. Weed control efficacy varies a lot in these reports. A study by Rasmussen (1992) recorded a weed control efficacy of 68.4 % with a spring tine harrow in peas and spring wheat. In oat, Rydberg (1994) measured a weed control success of 57-75 %, whereas Lötjönen and Mikkola (2000) found only 48-56 % and Rueda-Ayala *et al.* (2015) achieved an average of 51 % weed control efficacy in maize. Abundance of several weed species including grass weeds (*Alopecurus myosuroides* H., *Apera spica-venti* L., *Lolium multiflorum* LAM., *Poa annua* L.) and perennial weed species (*Cirsium arvense* L., *Elymus repens* L.) have increased in European cereal production systems during the last decades (Melander *et al.*, 2003). This was mainly due to higher percentages of winter crops in rotations and less intensive soil tillage operations (Massa *et al.*, 2013). Harrowing is less effective to control these type of weeds (Rueda-Ayala *et al.*, 2011). Therefore, hoeing might be a promising mechanical weeding method when higher weed control levels are required. Dierauer and Stöppler-Zimmer, H (1994) recorded 90 % of weeds between the crop rows, were uprooted and 75 % of the weeds within the crop row were covered by soil after two hoeing passes in maize and peas. While the weed control efficacy of uprooted weeds varies from 60 % to more than 90 %, the mortality of buried weeds can be poor depending on the amount of soil that is moved (Kurstjens and Kropff, 2001). Melander *et al.* (2003) recommended higher driving speed in order to move larger amounts of soil. Another strategy was the development of in-row tools to be able to selectively control weeds very close to crop plants or in between plants in the row. Mechanically driven finger weeders were already developed in the 1960s (Buddingh and Buddingh, 1963), and today are available for almost every hoeing equipment for wide-spaced crops (Kirchhoff and Duelks, 2019). A hoe equipped with inter-row sweeps and finger weeders can provide a very effective weed control success in just one run.

In vegetable cropping were plant distances are higher and driving speed is not needed to be fast, also powered tools that automatically hoe around every single crop plants are available (Dedousis *et al.*, 2007; Poulsen, 2018).

The seed row distance in cereals and the potential yield loss was also part of many studies. The recommended row distance varies between 180 and 250 mm (Lötjönen and Mikkola, 2000; Rasmussen, 2004; Melander, 2006). Other studies outline the higher yield potential in cereal crops that can be achieved by seed row distances smaller than 150 mm (Champion *et al.*, 1998; Boström *et al.*, 2012; Benaragama *et al.*, 2016). Lötjönen and Mikkola (2000), Mülle and Heege (1981) and Hakansson (1984) determined a yield loss of 12-15 % in spring barley when row spacing was increased from 125 to 250 mm.

To further enhance the guidance performance the first imaging sensor used for automated steering was introduced in the early '90s. At that time, also the first computer vision guidance systems were invented and tested (Tillett, 1991). Tillett *et al.* (2002); Nørremark *et al.* (2012) combined GNSS and optical sensors guidance to replace the poor performance of manual steering. Melander (2006); Griepentrog *et al.* (2006) showed that hoeing can be done faster and up to 40 mm close to the crop row. Even if the GNSS in combination with a locally provided RTK-correction signal is quite accurate, it does not meet the performance needed to further enhance hoe guidance precision.

For site-specific weed management strategies sophisticated systems also recording in the near-infrared area were used to distinguish between crop and weeds (Gerhards and Christensen, 2003). This is not necessary for a row guidance system where the discrimination is done just for plant material and background (Keicher and Seufert, 2000). Nowadays, using RGB-imaging sensors for precision guidance purposes is a common approach. They are relatively cheap and image processing can be also done online, as the consumption of computing power is low.

1.4 Automatic guidance

1.4.1 Advantages of automated row guidance of a hoe

To overcome the drawbacks of conventional hoeing machines, these systems today can be equipped with a precise and independent row guidance. When row guidance systems were introduced, they were only available for wide row spaces. Lately, some substantial evolvements of the technology enable the use in rather narrow row distance. The most obvious advantage of row guidance systems is the higher operating speed with high precision (Tillett and Hague, 1999; Tillett, 2005). Once adjusted for the specific field conditions, such systems operate stably over a long time and surface areas (Tillett, 2005). Compared to manually steered hoes, there is no need for a second operator. In contrast to center-mounted hoes, where the driver guides the tools by himself, an automated steered hoe does not demand high precision steering. Therefore once adjusted, also less experienced operators can run a hoe which creates more flexibility in labour organisation during work peaks. Higher operating speeds also allow adjusting the velocity according to the various soil conditions within one field. Due to the enhanced field efficiency, hoeing may become an option also for larger farms. Precise steering with row guidance systems allows to hoe as close as possible to the crop rows, and therefore reduces the strip where no direct treatments were manageable with manual guided equipment. A major drawback is that this systems can be quite expensive and as the technology is relatively new, further development is needed to improve the guidance stability especially for narrow row spaced crops.

1.4.2 Development of automatic guidance

Automated steering systems for agriculture vehicles were developed already 100 years ago (Wilson, 2000). First models had pre-applied furrows as a lead for wheels that provided mechanical steering (Phillips, 1949; Grovum and Zoerb, 1970; Kawasaki *et al.*, 1981). According to reviews from Jahns (1976, 1983), all these inventions were poor in the precision of guidance. Nowadays, the most promising solutions for autonomous steering are the Global Navigation Satellite System (GNSS) and machine vision-based approaches (Slaughter *et al.*, 2008; Mousazadeh, 2013).

1.4.3 Machine vision

Even though machine vision is very advanced for quality monitoring in industrial production, the environment in agriculture is a big challenge for algorithm development based on camera sensors (Hague *et al.*, 2000; Åstrand and Baerveldt, 2005). Changing light condition and varying growth stages, different soil types and moisture, as well as dust, lead to many different appearances of the same target object on images at different times (Tian and Slaughter, 1998; Hague *et al.*, 2000). Furthermore, the row as a target object can have gaps or consist of plants at different development stages. If weed infestation is at high levels, rows can be invisible (Åstrand and Baerveldt, 2005). To have a stable recognition of the target object, all possible parameter constellations need to be considered in the software (Tillett, 1991).

The processing part from capturing of images until providing a steering signal is therefore quite sophisticated and under permanently further development. At first, the camera image needs to be converted into a suitable digital colour-space for further processing. Next, the algorithm needs to discriminate per pixel between plant material and background material like soil, biomass residues and stones. A binary image can be created of the result where plant pixels are one class and background pixels are set as part of a second class. Different ways from the captured image to the binary image are described in the literature. One possibility is to use a Bayes-classification (Slaughter *et al.*, 1999), others used a threshold of colour saturation and brightness (Jia *et al.*, 1990). Later, threshold methods based on effective greenage index, as reported by (Woebbecke *et al.*, 1995) was used (Stanhope *et al.*, 2014). On the binary images the crop rows can be detect. In the '90s, when the first computer vision-based row guidance system were developed, a problem was the amount of captured data to process, which was limited by the computational power. Therefore, run-length encoding was used to implementing online machine vision for row guidance (Fehr and Gerrish, 1995) which lead to a poor row-recognising result. The most successful approach, to also address the problem of inconstant rows was based on the Hough-transformation (Hough, 1962). This method finds lines in images by recognising edge-pixels, passing multiple lines through that points and detect major intersection to find regular lines (Reid and Searcy, 1986). Marchant and Brivot (1995) showed an approach were the centre of crop plants where determined and the row pattern was detected based on a Hough transformation. To further stabilise the a proper steering signal, the use of a Kalmanfilter (Kalman,

1960) can help to estimate a row-recognition (Tillett *et al.*, 2002). This is done by mathematically combining variation of row appearances over multiple image and use the average information.

1.4.4   Row guidance for hoeing equipment

A set up for a machine vision steering systems for a hoe consists of an optical sensor, a controller unit and an actuator (Keicher and Seufert, 2000). An example sketch is shown in Figure 1.2.

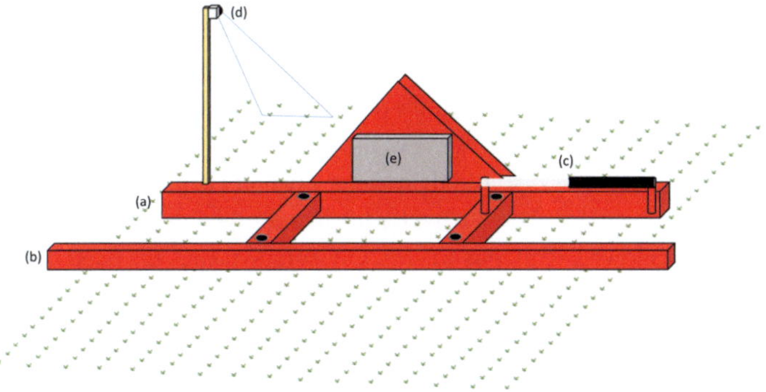

FIGURE 1.2: Camera steered hoe frame. **(a)** Stationary part, **(b)** Tool bar, **(c)** Hydraulic cylinder (actuator), **(d)** Optical sensor, **(e)** Controller unit. (B.Kollenda)

Suitable connection speed to have fast data transfers between the sensor and the attached controller unit is important (Tillett, 1991). Today, USB or Ethernet is typically used to connect a camera to a frame grabber unit. In most setups the sensor is mounted at the flexible part (e.g. the tool bar) moved by the hydraulic cylinder (actuator) (e.g. Slaughter *et al.* (1999); Tillett and Hague (1999)). A permanent tracking of the row and a possible offset of the implement generates a steering signal, which is executed as left steering, right steering or no movement. The signal is sent to the controller unit, where it will be translated into electrical power to trigger the solenoid valves for moving the actuator (Tillett and Hague, 1999). Additionally, a position signal, a speed signal, and a hitch position signal are sent to the controller or the embedded computer to adjust the guidance algorithm and put them into a break-mode during headland turns. The lateral steering of the hoe to correct the position of the tools is done mostly by hydraulic cylinders. The electric signal provided by the controller is converted into a hydraulic movement by the use of solenoid valves. In some cases just simple on-off valves are used while in more sophisticated setups, where also driving speed is tracked, a proportional pulse triggered valve is installed. Proportional valves can also be used to balance the working speed of differential cylinders. Some systems use the hydraulic pressure provided by the tractor and some models also generate the operating pressure independently with a PTO- driven hydraulic pump (E.g. Poulsen (2018)).

To use a camera sensor for row guidance, steering signals generated by an image processing algorithm need to correct the position of the hoe tools permanently. To get there, an option is to use hydraulic steered discs or cone-wheels in combination with loosened lower lift arms on the tractor (e.g. Tillett *et al.* (2008)). As in such a setup the width of movement is limited and the steering can be delayed or be indirect, so-called side shift systems, are more common. In these systems, a fixed part is mounted on the tractor hitch and a flexible tool bar is installed in a way that allows lateral movements. On some systems, the fixed part consists of a three-point hitch unit and a frame with a linear bearing system connects to the tool bar (ROW-GUARD, Einböck GmbH & CoKG). Another version of that side-shift system is a compact central mounted steering unit, with the possibility to install it between the tractor and an existing conventional hoe (Robocrop, Garford Farm Machinery Ltd. Other developments use a parallel steering unit where the flexible connection between the two frames is realised with two pivot arms (Slaughter *et al.*, 1999; Tillett and Hague, 1999). The guidance precision is the key factor to enable hoe technology in narrow seed rows and to introduce it on a wider scale in conventional agriculture. Today, there are plenty of camera guided hoe systems available on the European market. The most important systems are the Claas CULTI CAM (CLAAS E-Systems GmbH, Dissen am Teutoburger Wald, Germany), an OEM steering solution based on a Stereo RGB-camera to enable 3D-row recognition. The S300-Kit equipped with a bispectral sensor (Poulsen Engineering APS, Hvalsoe, Denmark). The Okio camera system consists of a set of two RGB-Cameras and a single row detection algorithm (Ensio GmbH, Braunau am Inn, Austria and Maschinenfabrik Maschienenfabrik Schmotzer GmbH, Bad Windsheim, Germany). A single RGB-camera steering system is manufactured by Tillett and Hague Technology Ltd. (Silsoe Beds, United Kingdom) and Steketee Camera IC-Light also based on a single RGB-Camera (Machinefabriek

Steketee B.V. Machinefabriek B.V., Stad aan 't Haringvliet, Netherlands). All mentioned hoeing systems were not available for row distances smaller than 180 mm. Thus, they cannot be used without further development in narrow seeded cereals. An important step for successful implementation of hoeing in narrow seeded cereals is a further developed row guidance that can recognise crop rows stable also at high weed infestation levels and at late growth stages were the crop leaves start to overlap.

1.5 Aims

The main goal of the presented study was to enable hoeing in narrow seeded cereals. As there was no solution available to hoe in crop rows smaller than 180 mm, a new concept was developed. For successful mechanical weeding in narrow seeded rows, a hoe tool needs to have a relatively compact design. The ideal tool-type should be able to reduce weeds at different growing stages, root or leaf shapes. Furthermore, the tool should work successfully under various soil conditions and different cropping systems without labour intensive readjustments. Beyond that, the tool and the arrangement of tool combinations should prevent clogging by biomass residues, soil chunks, or uprooted weeds. Finally, it also should work well under a wide range of driving speeds, with a uniform amount of soil disturbance and soil movement in the crop rows. Further on, a camera steering system is needed, which is able to detect the crop rows under various light conditions, growth stages and site-specific morphological differentiation. Also, different levels of weed infestation and densities should not influence the signal processing quality, to ensure stable and precise steering to guide the hoe through the rather narrow crop rows. The proposed hoeing system should be able to reach a weed control efficacy comparable to that of herbicide applications and simultaneously operate with a driving speed and field efficiency to compete for state-of-the-art spraying equipment. The specific objectives were:

1. Selecting and modifying potentially suitable hoe tools for the use in narrow seeded cereals.

2. Investigating the tools on proper weed control efficacy without damaging crop plants.

3. Testing of camera row guidance systems on their usability in narrow row distances to enable higher driving speeds without impacting negatively crop selectivity and weed control efficacy.

4. Designing and testing of prototype hoes with a combination of a camera row guidance system and the selected hoe tool.

5. Analysing the potential of hoeing as a weed management strategy in narrow seeded cereals alone and in combination with other mechanical and chemical weeding methods.

Chapter 2

Materials and Methods

This chapter contains information about all the materials and methods used in this study. The hoeing equipment is described in the first part. Secondly, the sites and conditions of the field trials are listed for every year (2.5). The analysis approaches are described in the last part of this Chapter (2.6). Specific information for every experiment is presented in Chapter 3.

2.1 Hoeing Equipment

If not specifically mentioned, all hoeing equipment was provided by Kress Umweltschonende Landtechnik (K.U.L.T., Vaihingen Enz, Germany). The assembling of the hoes was mainly done by the Department of Weed Science (University of Hohenheim, Germany) under supervision of K.U.L.T.. Only the first version of an hydraulic steered frame (see 2.5) was provided ready-to-use by K.U.L.T.. Tools which were not available in a suitable size for the requirements in this study were cut by an angle grinder. In cases of sharpened ends, the cut was re-sharped according to the original blade shape by the use of an angle grinder or a double grinder (Figure 2.1).

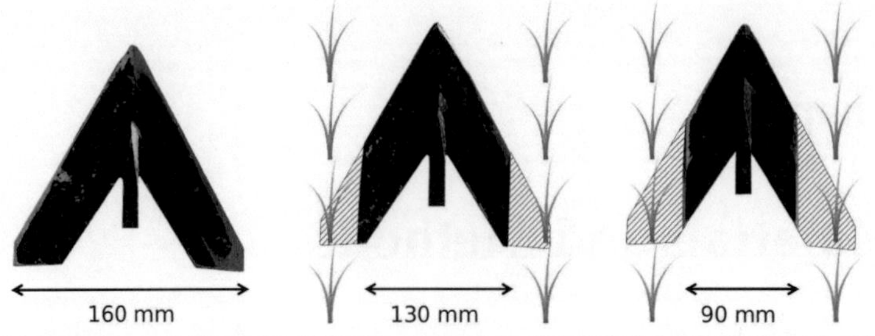

FIGURE 2.1: Adjustment of the tools for narrow row distances. The example is shown from left to right the available size (160 mm), the cut tool (130 mm) for the use in 150 mm row width, and the cut tool (90 mm) for the use in 125 mm row width (adapted from J. Machleb)

The used tool bars were recumbent H-shaped profiles to which parallelograms and further devices were easily attached by a screw tightened clip.

2.1.1 Parallelograms

To ensure a precise working depth of the tools in the soil surface, parallelograms were used in every hoeing-trial. Except for the first pre-test, the hoes were set up with one parallelogram per row. For the pre-tests, the model "Duo" was used while in the further conducted trials with a manually steered hoe, the model "Duo Solo" was used. Later, the model "Argus" was chosen to conduct large area field trials with an extra thin feeler wheel of 40 mm width (Figure 2.2). Both Duo parallelograms were originally developed for vegetable crops and was chosen because of multiple options to mount hoe tools. One parallelogram is equipped with two feeler wheels and is used for two inter-row spaces. The Duo-Solo is a modified version of the Duo model: It is made for single inter-row spaces and one feeler-wheel is mounted in the centre. The Argus parallelogram is a more stable construction for crop cultivation under heavy soil conditions. It has an S-shape spring tool attachment.

Duo   Duo-Solo   Argus

FIGURE 2.2: Parallelogram models used in the study. All devices were provided by K.U.L.T. (B.Kollenda)

2.1.2 Selection of hoe tools

In Figure 2.3 the selection of tools for the first trials is presented. Tools are categorised as preparing tools, main tools and secondary tools. The preparing tools were designed to support the main tools by dividing the topsoil near the crop plants. This is done to prevent crop damage by avoiding breaking up larger soil chunks. Additionally, they pre-cut biomass residues to reduce the risk of clogging of the main tool. The main tools were selected to do the weeding by uprooting of weeds in the intrarow space. Secondary tools were installed behind the main tool to divide soil conglomerations and to uncover the roots of the weeds to dry them out.

Preparing tools

The hollow discs are common tools in many tillage equipment. The one used in that project had a diameter of 160 mm. They were used pairwise in the outer sides of the inter-row space and shaped concave facing towards the crop rows.

Main tools and secondary rake tool

The dam-shape knives were angled in the middle to nearly 90° and adjusted to have a horizontal part working under the surface soil. They were also used pairwise by driving the vertical part along the row and the horizontal part facing towards the middle of the intrarow space. They were slightly bent against the driving direction. While the discs move small amounts of soil from the crop

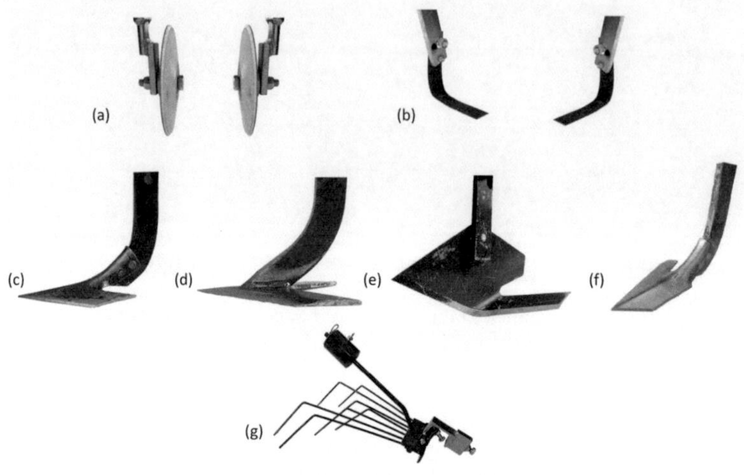

FIGURE 2.3: Selection of hoe tools. Preparing tools: **(a)** Hollow discs, **(b)** Dam-shape knife. Main tools: **(c)** Goosefoot sweeps, **(d)** No-till sweeps, **(e)** Down-cut side knife, **(f)** Half-goose foot. Secondary tool: **(g)** Rake element. (B.Kollenda)

row border into the intrarow scace, the modified dam-shape knifes side-cut and undercut the soil which results in topsoil disconnected from the crop row. The second main tool is a common and widely used hoe sweep. It consists of a goosefoot shaped flat steel with a curved crown along the centre. The original 160 mm working width was shortened down on both sides to 120 mm and 80 mm for 150 mm and 125 mm row spacing respectively. The second tool had a similar shape with a non-curved flat A-shape steel knife with sharped blades on the outer sides. It was modified for 150 mm row distance by cutting the sides to 130 mm and a 125 mm tool was cut to 90 mm. The down-cut side knife blade was originally developed for mechanical weeding in vegetable crops. It combines preparing tool characteristics with main tool features. It consists of a vertical driven diamond-shaped flat steel device with a blade at the front end faced towards the ground. This part cuts along the crop rows similar to the discs. An overview of the modification of hoe tools is given in Table 2.1.

A second knife, facing backwards and angled in driving direction, performs the weeding in the intrarow space. This tool was also used pairwise on both

TABLE 2.1: Overview of the modifications of the tools for the specific crop row distance. HD: Hoolow discs, DK: Dam-shape knife, GFS: Goose foot sweep, NTS: No-till sweep, DSK: Down-cut side knife, HGS: Half-goose foot sweep, RE: Rake element

Characteristics	Adaption for seed row with		
of original tools	200 mm	150 mm	125 mm
HD 160 mm	160 mm	160 mm	160 mm
DK 45°	90°	90°	90°
GFS 160 mm	160 mm	120 mm	80 mm
NTS 160 mm	160 mm	130 mm	90 mm
DSK 180 mm	160 mm	90 mm	65 mm
HGS 160 mm	20 mm one side	20 mm one side	10 mm one side
RE 100-300 mm	300 mm	140 mm	115 mm

outer sides of the intrarow space. Adjustments for smaller row distances were done similar to the dam-shape knives. Another tested tool was a modified goosefoot sweep, where the complete right or left part of the goosefoot was cut off. This tool was also used pairwise with the cut sides towards the crop rows. As a secondary hoe tool, an adjustable rake element was tested. The element was mountable at the tool handle of the main tools facing backwards. It consists of seven harrow tines and an adjustable mass to reduce or enlarge the raking force of the tines. The working width can be modified by spreading the tines in a lateral direction.

2.1.3 Manually steered hoe frame

A manually steered hoe frame model "Argus", (K.U.L.T.) with a track width of 1.5 m was used for the hoeing tool tests (Figure 2.4).

The hoe was mounted on the three-point hitch at the back of a tractor. The hoe frame was steered by an operator sitting on the back of the hoe. On the tractor tracks, parallelograms type "Argus" (K.U.L.T.) were used on both sides equipped with wider goosefoot sweeps shaped to fit tractor tire width.

2.1.4 Hydraulic steered Hoe Frame

To test the two camera systems a three-meter hydraulic steered hoe frame was provided by K.U.L.T.. The hoe frame was also a model "Argus" frame like the one used for the trials described above. The hoe was a parallel pivot arm

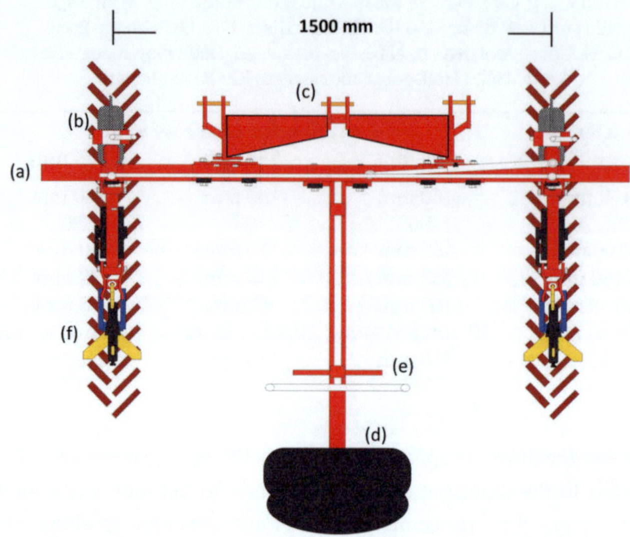

FIGURE 2.4: Schematic overview of the manual steered hoe frame. **(a)**, **(b)** steering wheels, **(c)** three-point hitch connector, **(d)** operator seat, **(e)** handlebar, **(f)** 160 mm goosefoot sweeps for the tractor track. (Kress Umweltschonende Landtechnik, K.U.L.T., Vaihingen Enz, Germany)

steering system with a centrally mounted synchronous cylinder with 400 mm working way (Chapel Hydraulique GmbH, Günzburg, Germany). To track the movement of the hydraulic cylinder, a 450 mm potentiometer cylinder model "PC-67" (Gefran GmbH, Seligenstadt, Germany) was mounted in parallel, to send the current position signal to the controller unit. The hydraulic cylinder was connected to the flexible tool bar of the frame via a 28 mm diameter tie rod. Two pivot-arms were linked to the tool bar and stationary part. A third tower arm construction linked the upper hitch point with the central part of the tool bar. As a three-meter hoe have not enough distance to the tractor left and right the camera was installed under the tower arm looking in between steered and stationary tool bar (Figure 2.5).

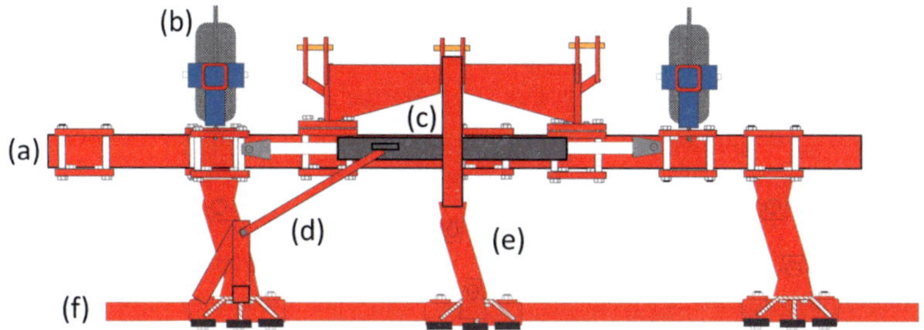

FIGURE 2.5: Hydraulic steering frame. **(a)** Stationary part **(b)** Supporting wheels, **(c)** Synchronous cylinder, **(d)** Tie rod, **(e)** Tower arm, **(f)** Tool bar (B.Kollenda)

2.2 Camera row guidance equipment and the test setup

One important milestone of the project was to enable camera steering of a hoe in narrow row distances, smaller than 180 mm. A system needed to be found that can be further developed and adjusted. The requirements on the system were:

1. A Camera sensor in a dust and water sealed case that is mountable on the movable tool bar of the hoe.

2. A controller unit that provides a steering signal calculated from an offset of the camera view and the crop rows between 250 mm and 120 mm.

3. An interface to enter specifications for the hoe operation and stream the camera field of view in real-time for supervision.

The idea was to modify and fine-tune an already available system for the specific precision needed in this study.

2.2.1 Naïo Robot Camera System

The first tests were done with a modified camera system originally developed for a robot-system (Naïo Technologies, Escalquens, France). To calibrate the software for the new task, images were captured in winter wheat field sown with a row distance of 150 mm, during a typical growth stages were hoeing should be performed (Figure 2.6). The system consists of a camera model BlueFox IGC200wc (MATRIX VISION GmbH, Oppenweiler, Germany), with a CMOS sensor and a resolution of 752x480 pixels. Images were processed on an Intel Celeron® based embedded system. The system provides the steering signal calculated by the image analysis, that is transferred as analogue electrical signals between 0.1 to 10 V (where 5 V is the "centre position" signal). The system was provided without a user interface. Therefore, a capacitive 7-inch display was connected to the display port of the embedded system. As the code was running under LINUX environment a graphical version of Ubuntu 16.04 was installed to run the software inside a graphical session. After that, it was easy during testing and operating, to change the adjustments in the configuration file pre-defining the camera angle, the number of rows in scope and the camera height .

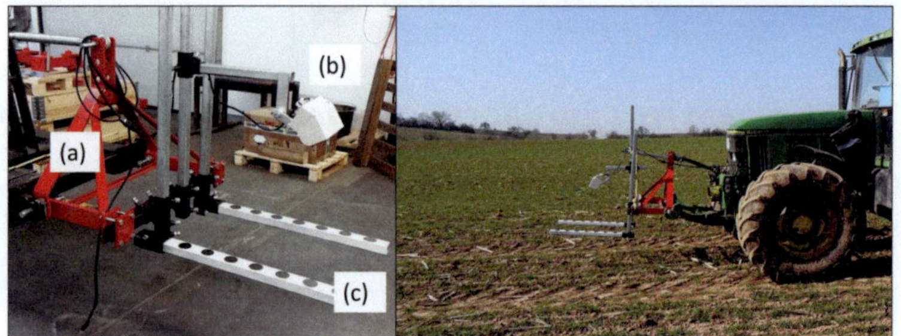

FIGURE 2.6: Reference image capturing for software adjustments at the Naïo guidance system. Left: **(a)** Mounting frame, **(b)** Camera, **(c)** Reference bars. Right: Capture drive in winter wheat (B.Kollenda)

2.2.2 Tillett and Hague Camera System

A second strategy uses a commercially available camera system Inter-Row Vision Control from Tillett and Hague Technology Ltd (Silsoe, United Kingdom). The set includes an Ethernet RGB-Camera with a CMOS sensor model Blackfly® from FLIR Systems (Wilsonville, US) and a controller Unit with an output signal for solenoid valves. The unit has also an input for a position sensor and an odometer signal. Furthermore, there is an input for a lift-up sensor. The image processing is done on an embedded system inside a user interface with a 7-inch display. No information was available about the type of the computer system.

2.2.3   Set up to test the Naïo camera system

An overview of the hoe equiped with the Naïo camera system is given in Figure 2.7. The cylinder was moved by a directional solenoid control valves with a proportional magnet.

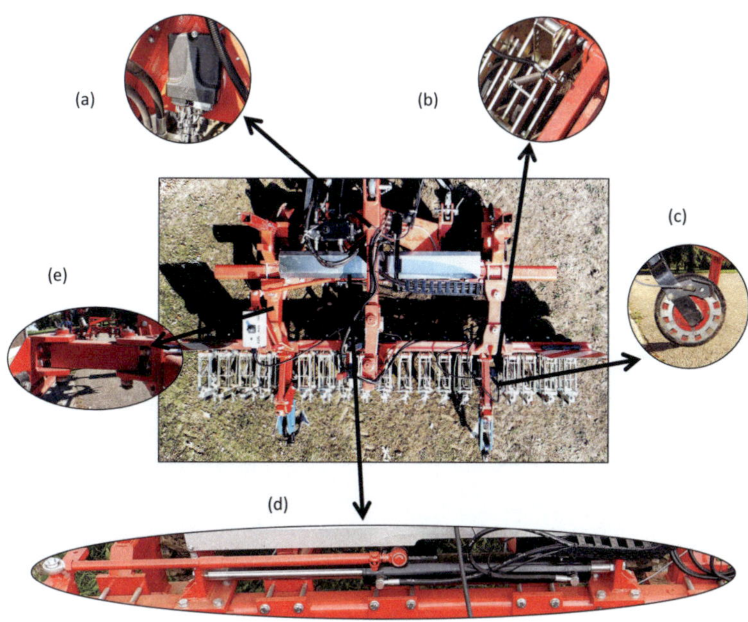

FIGURE 2.7: Hoe used to test the Naïo camera system. **(a)** Hydraforce controller "valve driver", **(b)** lifting sensor, **(c)** Reluctance sensor, **(d)** Synchronous cylinder, **(e)** Pivot hinge (B.Kollenda)

Valve type was a "DSE3" from Duplomatic (Duplomatic MS S.p.A., Parabiago, Italy). The valve was mounted on a hydraulic distribution block where the pressure hoses from the tractor were connected to the hydraulic steering cylinder. There was also a connector for a pressure gauge. To enable specific adjustments and extra modes, a controller model "ECDR Valve driver" (Hydraforce, Lincolnshire, IL USA) was installed between the output of the camera controller and the solenoid valves. At one parallelogram a reluctance sensor was mounted on the wheel where a punched disc mounted in the rim was used as the reluctor ring. Another reluctance sensor was mounted to give a signal if the hoe is lifted. It was placed on an arm at the side of a parallelogram swing in the position that matches the swing in the lifted position of the hoe. A remote control allows to override the camera signal and to shift the lateral position manually.

2.2.4 Set up to test the Tillett and Hague camera row guidance

To use the second camera system, the steering system was modified in cooperation with K.U.L.T. as described below. The Hydraforce controller was removed and the steering signal provided from the Tillet and Hague system was connected directly with the solenoid valve. The hydraulic system was changed to an on/off based operation: The valve was a directional solenoid valve type "Cetop" (Kramp GmbH, Strullendorf, Germany). The other items remained unchanged (Figure 2.8).

FIGURE 2.8: Setup to test the Tillett and Hague row guidance. **(a)** RGB-Camera, **(b)** Control unit, **(c)** Solenoid Valves and hydraulic block, **(d)** Interface and embedded system, **(e)** Remote control. (B.Kollenda)

2.3 Three-meter hoe prototype

The first prototype of the camera steered hoe was built with a three-meter work width, equipment of the Tillett and Hague setup (Figure 2.9). This was done as most sowing operations on farms in Southern Germany is done at the same width. So, the setup of the field trials could be done in line with local agronomic practices. Furthermore, the hoe can be transported more easily without cumbersome dismantling or remounting. As the Naïo Camera system was not reliable enough during the pre-tests, the decision was made to use the Tillett and Hague row guidance for the final hoe prototype.

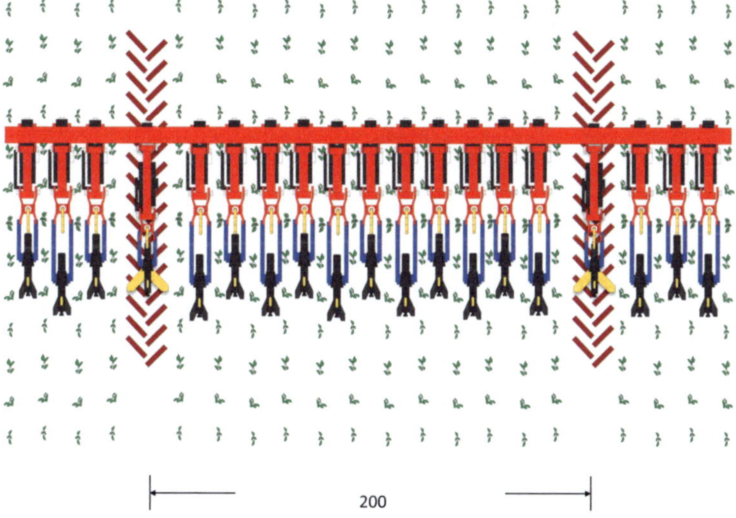

FIGURE 2.9: Hoe tool arrangement of the 3 m camera guided hoe.
(B.Kollenda)

2.3.1 Construction details

The three-point hitch frame and all the pivot arm steering parts were taken from the hydraulic test setup described above. The hydraulic system was equipped with two manual flow-control valves "FT 257/5" (F.lli Tognella Spa, Somma Lombardo, Italy) to fine-tune the oil pressure provided by the tractor (Figure 2.10). The valves were attached between the solenoid valves and the hydraulic cylinder. The Tillett and Hague controller unit was mounted under the three-point hitch frame to protect the unit from damage during operations and transport. The camera was mounted below the centre pivot arm of the tower. As parallelograms, the model "Argus" was used for every row. The positions of the equipment enabled accurate tool guidance. Alternating short and long parallelogram endings enlarged the distance between the tools. No-till sweeps with 75 mm were used

in every row. 160 mm no-till sweeps were mounted to operate in the driving
track of the tractor wheels.

FIGURE 2.10: Construction details of the 3 m camera guided hoe.
(a) Tillett and Hague controller unit, **(b)** Flow-control valves, **(c)**
Solenoid valves. (B.Kollenda)

2.4   6 m segmented hoe prototype

As high area performance is a key factor of successful weed control by hoeing,
the working width of the hoe is a major bottleneck to compete with herbicide
applications. As the optimum speed to hoe is limited by too much soil movement
and increased hoe hopping during operations, enlarging the width was the next
step in the optimisation of field efficiency. At the same time, sowing at three-
meter row width impeded to just enlarge the hoe frame (Figure 1.1). Further on,
the wider working width bears the risk of a decline in hoeing precision due to the
increased tool movements of the whole hoe setup. In narrow seeded cereals, this
problem can lead to crop damage on the outer side tools of the hoe, even when
the centre part is guided correctly in the crop rows. One way to overcome all
these aspects is to split the steered parts of the hoe into two independent devices.
Consequently, a concept was developed for a hoe with a 6 m working width
separated into two three-meter elements. Each element has an independent
camera and a separate hydraulic steering.

2.4.1   Folding frame

The proposed hoe concept needed to meet legal requirements for driving on
public roads in Germany. The maximum allowed dimensions of an agricultural

machine mounted at the three-point hitch are three-meter wide by four-meter high. Therefore, the hoe needed to be foldable. A 6.6 m folding frame divided in three 2.2 m elements were used at the basis of the hoe. The frame consists of a 125 mm steel square pipe and a 110 m steel square pipe parallel to each other. Material thickness was 4 mm and both devices were connected every 1.1 m by flat steel elements with 5 mm thickness. The two outer elements were connected by hinges with the middle element. Double action cylinders enable folding to a U-shape of 2.4 m width (Figure 2.11). At the middle element, there was a trapeze-shape construction used to mount the frame at the three-point hitch. As the original mounting points were too close to the frame, to also mount supporting wheels, a three-point hitch frame was attached at the original mounting points to enlarge the distance between frame and tractor.

FIGURE 2.11: Folding frame for the 6 m hoe. The left picture show
the unfolded frame and the right picture show the frame folded.
(B.Kollenda)

2.4.2 Pivot arm steering

Pivot arm steering elements and a middle tower pivot arm were attached to the outer elements on both sides to have three linkage points per side. Two three-meter tool bars were connected to a 700 mm toolbar end by a hinge to be able to move the centre parts of the tool bar away to allow the folding of the basic frame. If both tool bars were centred the inner gap in between the two bars is 420 mm wide. The steering rods and the actuators (hydraulic cylinders) were on the outside pivot arms on each side.

2.4.3 Camera system

A double row guidance system (Tilled and Hague Inter-Row Vision Control) that consisted of two cameras, two controller units connected to each other and an interface was used. The interface can process the images from both cameras simultaneously and provide the field of view from both cameras, by manually switching between the two streams of the cameras. The cameras were mounted in front of the hoe frame on L-shaped support arms made of 45 mm square pipe steel. The camera support arms were mounted on the top of each tool bar above the foldable frame. The cameras field of view was left and right of the tractor inside the two three-meter sowing strips.

2.4.4 Hydraulic system

The hydraulic system of the steering was completely separated from the circuit used for the folding of the basic frame. The system consists of the hydraulic distribution unit and two hydraulic cylinders each for one section of the hoe (Figure 2.12).

FIGURE 2.12: Construction details of the 6 m camera guided hoe. **(a)** Control units for each segment, **(b)** Input of hydraulic pressure from the tractor with a debris filter, **(c)** Hydraulic return flow to the tractor, **(d)** Bypass return flow, **(e)** Adjustable pressure balance, **(f)** Solenoid valves. (B.Kollenda)

The cylinders were double-action cylinders with 400 mm total way (Lind Jensens Maskinfabrik A/S, Højmark, Denmark). The hydraulic distribution unit consist of a foot blade, where solenoid valves with a proportional magnet were mounted. Valve type was also a "DSE3" from Duplomatic (Duplomatic MS S.p.A., Parabiago, Italy).

2.4.5 Anti-collision system

A major challenge for the design was caused by the potentially overlapping of the movement range of the tool bars. If one side accidentally overrides more than two rows by an error of the camera row guidance, the two bars can collide. Avoiding that problem by mounting the tool bars behind each other was not an option because the maximum equipment length to drive on public roads was already reached. Further on the increased leverage would have been beyond the capacity of the three-point hitch. The solution was to develop a sensor-based anti-collision system. An ultrasonic sensor was used to monitor the distance of the two tool bar endings. A 300 mm by 160 mm metal reflection plate was mounted at the of the left tool bar and the sensor was placed on the right tool bar end, facing towards the metal plate (Figure 2.13).

FIGURE 2.13: Anti-collision system of the 6 m segmented hoe. **(a)** Ultrasonic sensor, **(b)** Reflection plate. (B.Kollenda)

To interrupt the steering signal send from the Tillett and Hague system to the solenoid valves, a programmable input-output micro controller was connected. The controller model was an ecomat Mobile Basic CR0403 (ifm electronic GmbH, Essen, Germany) with 12 digital outputs, 12 digital inputs, as well as 4 analogue inputs. The controller was programmed by K.U.L.T.. The Tillet and Hague system was connected to the digital inputs. The sensor was able to create a pulse width modulated (PWM) output signal which enabled the solenoid valves to have different flow rates. In standard operation mode, the controller provides an output signal to the solenoid valves whenever there is an incoming signal from the Tillet and Hague system (Figure 2.14). In case the ultrasonic sensor is triggered by a too close position of the tool bars the inside moving signals of both sides were blocked. Further on an alarm sound informs the operator about the issue.

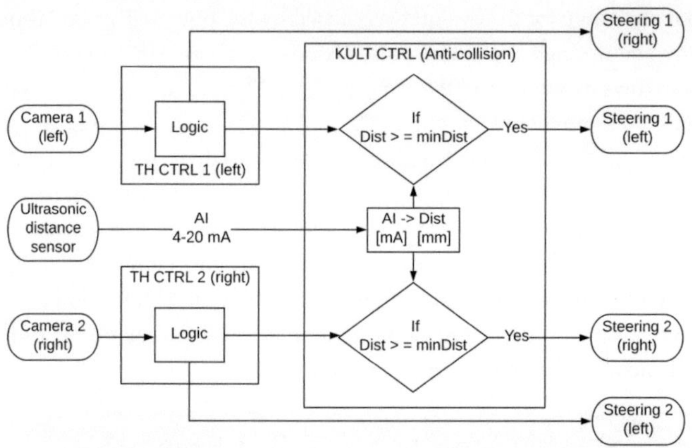

FIGURE 2.14: Diagram of the anti-collision system of the 6 m segmented hoe. **TH CTRL:** Tillett and Hague controller, **KULT CTRL:** ecomat Mobile Basic CR0403 (adopted from D. Riehle)

2.4.6 Tool equipment

For testing the hoe in narrow seeded cereals, the prototype was equipped with 36 parallelograms distributed on each side. To have the same accuracy of depth leading already tested with the 3 m model, the 6 m hoe was equipped with one parallelogram per row. Also, the model "Argus" equipped with the 40 mm wheels was used. In the centre tools were attached in alternating positions to allow close approximation of both tool bars. On the left side, an enlarged version of the parallelogram was installed to have the tools out of the area where the left parallelogram is working. The inter-row space in the centre was treated by a single more narrow cut no-till sweep mounted at an overhanging arm connected to the long parallelogram end (Figure 2.15).

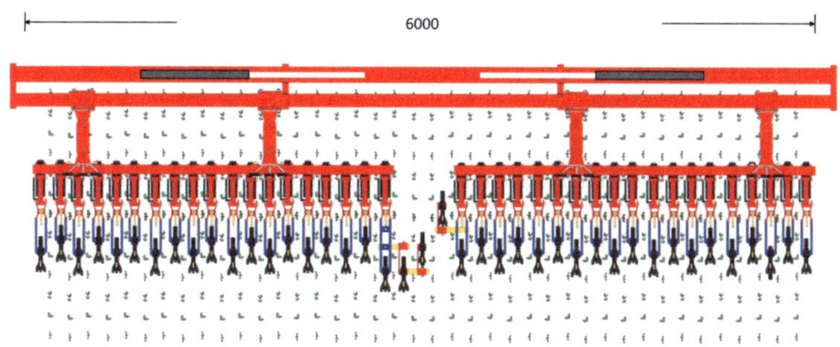

FIGURE 2.15: Concept of the segmented 6 m hoe and the tool
arrangement for 150 mm seeded cerels. (B.Kollenda)

2.5 Field trials

The field trials described in this study were conducted at two different trial
stations of the University of Hohenheim (South Germany). One is located near
the campus of the University (HOH). The other one is located near Renningen
(Germany) at Ihinger Hof (IHO). Plot seeders were used to establish the trial
sites. A customised plot seeder provided by Deppe (Agrar-Markt Deppe, Bad
Lauterberg-Barbis, Germany) was used at IHO and a Hege plot seeder (Zürn
Harvesting, Schöntal-Westernhausen, Germany) at HOH. Both seeders were
equipped with double disc coulters (Lemken, Alpen, Germany) to achieve a
proper paralleling of the rows to each other. If not other mentioned seeding
equipment was adjusted to place seeds in 30 mm depth.
The soil type is a clayey loam on both research stations. 2017 was the rainiest year
for HOH with 830 mm rainfall, compared to the long-term average of 647 mm. In
march where treatments were conducted, there was a slot with no rain. At IHO
the total measured rainfall summed up to 653 mm in 2017 which was 80 mm
less than the long-term average. In March, there were also weather conditions
with no rain that allows hoeing. Compare to HOH, at IHO was also less rainfall
during the rest of the growing period. Mean temperatures in 2017 at both sites
were close to the long-term average with 10.2 °C at HOH and 9.2 °C at IHO
compare to HOH (Figure 2.16).

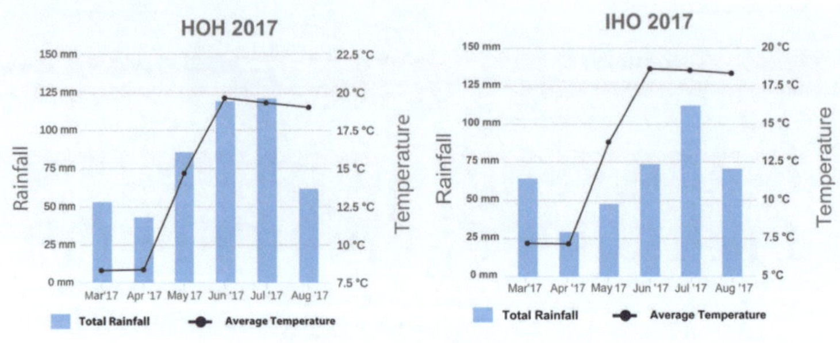

FIGURE 2.16: Weather conditions of the 2017 season. (Graphs: Agrarmeterologie BW)

In 2018 the total rain was 525 mm in 2018 for HOH and 649 mm at IHO, which marks a very dry year compared to the long-term average at both trial sites. Hoeing was easy to conduct under suitable weather conditions. At HOH, the month with the highest rainfall was June while at IHO there was a lot of rain in Mai and more dry weather conditions in the following months. Temperatures in spring were rising very fast. Compared to 2017 the average was higher than 12 °C from the end of March onwards at both sites (Figure 2.17).

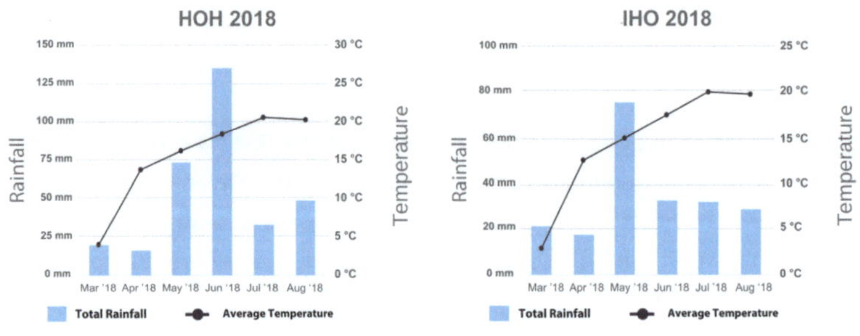

FIGURE 2.17: Weather conditions of the 2018 season. (Graphs: Agrarmeterologie BW)

In 2019 at HOH, in February there was only half the rainfall and 3.2 °C warmer compared to the long-term average records. At both sites also in Mai a lot of rain events were recorded. The total rainfall between February and August 2019 was 366,9 mm at HOH and 502 mm at IHO (Figure 2.18).

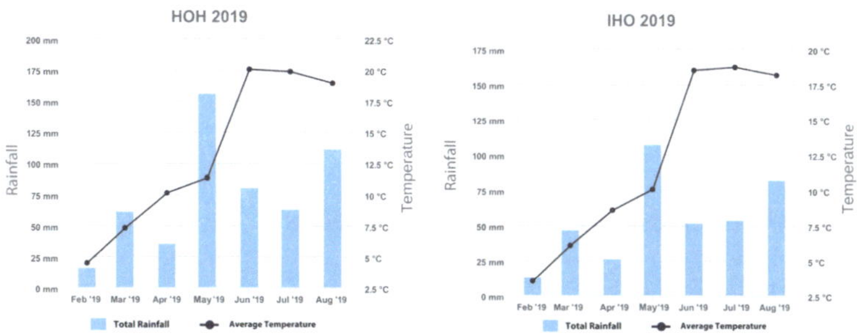

FIGURE 2.18: Weather conditions of the 2019 season. (Graphs: Agrarmeterologie BW)

2.5.1 Trial design

All trials were set up as a randomised complete block design with different hoeing treatments as the experimental factor. The treatments were randomised for each repetition. If not mentioned otherwise, each treatment was replicated four times.

Plot trials were set up at least 10 m inside the field from each side. Further on, the trials were aligned to achieve the most possible even soil structure along with the treatments according to the research stations experience. In the case of sloped sites, repetitions were aligned along the gradient of the slope.

2.5.2 Field test bed

A field test bed consisting of some test strips were prepared to check the camera steering under realistic operation parameters, with most environmental influences. Wheat was sown in 1.5 m long strips and 150 mm row distance. Tests were conducted with wheat at four-leaf stage or later. If later, plants were cut down by a hand-driven lawn mower and plant residues were removed. A part of the field test bed area was hand weeded, while other parts were left weedy, to investigate the performance of the camera systems under various condition.

2.5.3 Herbicide application

All herbicides used in the trial are described in Chapter 3 inside the specific section to each experiment. The application was done with a self-propelled plot sprayer from Schachtner (Ludwigsburg, Germany). Nozzle type was an IDK Flatfan from Lechler (Lechler GmbH, Metzingen, Germany). The pressure was set to 300 kpa at 200 l ha^{-1} of spray volume. The sprayer was equipped with a variable boom that allows application widths from 2 to 3 m.

2.6 Methodology of measurements

2.6.1 Weed counting

Weed counting was done for every trial before hoeing and right afterwards. To record new emerged and survived weeds another counting was done fourteen days after the treatments. A 33 x 33 cm frame was randomly positioned at three locations per plot. All weeds inside the frame were counted separated by species. The assessments were done separately for inter- and intrarow space.

2.6.2 Crop soil cover determination

To determine the efficacy of weed control inside the row and the corresponding selectivity of treatment to the crop, a method originally introduced by Rasmussen and Svenningsen (1995), where crop leaf cover with soil, the crop soil cover (CSC) is assessed. The approach was further developed and tested in winter wheat. The method is based on digital image capturing and an automated image processing algorithm.

Sensors

Two different cameras were used to capture the images. The first was a handheld camera, specifically Sony DSC-HX60V (Sony Corp., Minato, Tokyo, Japan). The camera was equipped with an Exmor R® CMOS sensor (Sony Corp., Minato, Tokyo, Japan) with a diagonal sensor size of 7.76 mm and an image resolution of 20.4 million pixels. Images were taken with a shutter of 1/640 s and a focal length of 4.3 mm. In order to achieve the necessary stability of the images, a 2 s delay was chosen between triggering the camera and the actual capture, giving time to the operator to standoff. The camera was mounted on a tripod with a distance of 600 mm to the ground. The same height was used to capture images by a smartphone model SM-G930F (Samsung Electronics Co., Suwon, South Korea). The later was equipped with an IMX260 CMOS sensor (Sony Corp.,

Minato, Tokyo, Japan) with a diagonal sensor size of 10,6 mm and an image resolution of 12 million pixels. Images were taken with a shutter of 1/236 s and a focal length of 4.2 mm. The proprietary software, provided by the manufacturer, was used in each sensor to capture and store the images.

Segmentation Algorithm

For the image processing, a customised algorithm was created to process the images. Each RGB image was transformed in the hue-saturation-value (HSV) color space 2.19. In this colour space, a bandpass filter was applied to remove all outlier colours and the majority of the soil chucks. This was achieved by removing all continuous components above a specific size. These were the components that belonged to soil chucks. The combination of that two information was utilised to create a gross background mask that removed the majority of non-plant material.

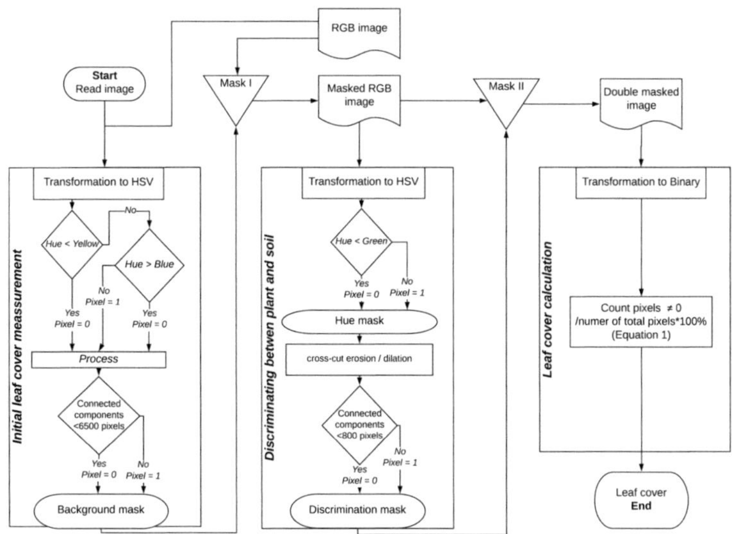

FIGURE 2.19: Flowchart of the script. On the left side based on the original image, a first mask is calculated to hide soil areas. In the middle, a second fine mask is calculated and overlaid on the image to segment plant material. Further on, at the right side the leaf cover is calculated in percentage (See Equation 1). (B.Kollenda)

In this masked image a second and finer filtering was performed. In the HSV color space, a low pass filter in the green region was used to remove remaining soil material. Furthermore, a cross-cut erosion/dilation was performed. Small noise regions that were re-included in the image from the previous step were removed by deleting big components. The overlay of both masks was applied on the original image in order to segment between plant material and non-plant material. This overlaid mask can simultaneously provide information about the leaf cover of the image by counting the ratio of zero and non-zero pixels as described in Equation 1:

$$\text{LC} = \frac{\sum px_1}{\sum px_t} \times 100 \ \%$$

(1)

2.6.3   Grain yield determination

Crop yield was recorded by harvesting the trials with a plot harvester. Plots were adjusted to a size where no side-effects influence the results. Therefore, the beginning and the end of every plot was shortened to a standard plot length. Additionally, just a core section in the center of the original plot was harvested. Grain samples were taken from every plot to measure the moisture of the grain. The yield was determined for 14 % grain moisture in each plot.

2.6.4   Tacheometer records

In order to determine the accuracy of the camera row guidance system in combination with the hydraulic steering, a record of the camera steered hoe was done by an automatic tacheometer model "STS930" (Trimble, Raunheim, Germany). The tacheometer permanently measure the distance and the angle from the station to a prism, based on a reflected beam of a light. The distances were recorded as Cartesian coordinates for further analysis. The measurement accuracy was +/- 2 mm.

2.6.5   Data analysis

Weed control efficacy (WCE), was measured immediately after treatment according to Equation 1 originally proposed by Rasmussen (1991).

$$\text{WCE} = 100 - \frac{wds}{0.01 \times wdu}$$

(1)

where wds is the density of survived weeds in treated plots and wdu is the density of weeds in untreated control plots. Weed composition was recorded 14 days after treatment.

Crop soil cover was calculated according to Equation 2. It is the percentage of leaf area covered by soil directly after hoeing (Rasmussen *et al.*, 2008). L_t is the leaf cover in treated plots and L_m is the leaf area in the manual treated control plots.

$$CSC = 100(1 - L_t/L_m) \tag{2}$$

The selectivity was calculated by the ratio of weed control efficacy and crop soil cover according to Equation 3.

$$S = WCE/CSC \tag{3}$$

where S is the selectivity of the same treatments and WCE is the weed control efficacy. CSC is the crop soil cover (Equation 2) of the treatment.

2.6.6 Statistical analyses

The statistical analysis were performed with the R Studio software (Version 1.2.1335, RStudio Team, Boston, MA, USA). Before analysis, each data set was checked for normal distribution of the residuals and homogeneity of variance. After analysis of variance means, the observations were compared by a Duncan's multiple range test at a level of $\alpha \leq 0.05$. For each one factorial trial Model 1 was used:

$$y_{ij} = \mu + \beta_j + \tau_i + \epsilon_{ij} \tag{1}$$

where y_{ij} is the observation of the ith treatment in the jth block. μ is the general mean, β is the effect of the block j, τ is the effect of the treatment i. ϵ is the residual error.

If results from more than one trial site were analysed Model 2 was used:

$$y_{hij} = \mu + \alpha_h + \beta_i + \tau_{hi} + (\alpha\,\beta)_{hi} + \epsilon_{hij} \tag{2}$$

where y_{hij} is the observation at the trial site h of the ith treatment in the jth block, μ is the general mean, α is the effect of the trial site h, β is the effect of treatment i, τ is the effect of the block at trial side h and treatment i and $(\alpha\,\beta)_{hi}$ is the interaction of trial side h and treatment i. ϵ is the residual error.

Chapter 3

Details of the Experiments

3.1 Overview of the experiments

To further develop a hoe as a new weed management solution for narrow seeded cereals the experiments seen in Table 3.1 were conducted.

TABLE 3.1: Overview of all trials conducted for this study. **CSC**: Crop soil cover, **WM**: Weed management. *Test was done in maize due to the time of the season and availability of test area.

Experiment ID	Test	Row distance
1	CSC determination	125-200 mm
2	Pre-test of tools	125-200 mm
3	Test of selected tools	125-200 mm
4a	Row guidance adjustments	150 mm
4b	Accuracy test	750* mm
5	Test of the prototype hoes	150 mm
6	Integrated WM	150 mm

The success of a hoeing treatment was determined by the weed control efficacy, in combination with the amount of crop influence due to the weeding. While the first aspect was recorded by a weed species count per plot, the measurement of crop soil cover (CSC) as proposed in 2.6.2 first need to be tested and adjusted in Experiment 1. After the measurement methods are setup a selection of 12 tools and tool combinations was tested in Experiment 2. For the most promising tools a two years test series on different sites was conducted (Experiment 3). Additionally, in Experiment 4, two camera row guidance systems were investigated on their ability to perform at row spaces of 150 mm.

Prototype hoes, one with 3 m and another one with 6 m (the later with a double guidance system for each 3 m strip), were build. Afterwards, an accuracy test of both hoe-prototypes with a tacheometer was conducted. When both prototypes proofed to be precise, they were tested in Experiment 5 in field trials. Experiment 6 was also a two years trial series proofing to use hoeing in narrow row width in combination with a spring-tine harrow and a herbicide application.

3.2 Details of Experiment 1: Crop soil cover measurements

To adjust the new CSC measurement algorithm, a set of ten images was chosen. The images differed widely in terms of light conditions, amount of plant material and soil colour. After that, a set of 14 widely varying images was processed by the algorithm. The resulted masked images were compared visually with the original images. When the algorithm was advanced enough to deliver reliable results, a set of six images was hand segmented with a graphical image processing software (GNU Image Manipulation Program 2.8.22). For every hand segmented image, the CSC was calculated similarly to the approach in the automated algorithm. Afterwards, the unsegmented six images were also processed by the new algorithm and both CSC results were compared. All images were recorded in winter wheat field trials before and after treatment in fall 2016 and spring 2017 to cover the same situations as under trial conditions.

3.3 Details Experiment 2: Pre-test of tools

For testing the modified hoe tools and the combinations with the primary and secondary tools, a field trial was conducted in spring 2017. Every hoe treatment was tested between 3 and 4 km h^{-1}. An overview of the different treatments is given in Table 3.2.

TABLE 3.2: Overview of the treatments of the tool tests

Treatment	Weeding method
Trt. 1	Untreated control
Trt. 2	Permanent weed free
Trt. 3	Goose foot sweep
Trt. 4	No-till sweep
Trt. 5	Down-cut side knife + discs
Trt. 6	Half goose foot sweeps
Trt. 7	Dam-shape knife
Trt. 8	Goose foot sweep + rake
Trt. 9	No-till sweep + rake
Trt. 10	Down-cut side knife + rake
Trt. 11	Half goose foot sweep + rake
Trt. 12	Dam-shape knife + rake
Trt. 13	Discs
Trt. 14	Down-cut side knife

The main goal was to identify the tools, which have the best combination of weed control efficacy, handling and suitability over a wide range of soil conditions. The first trial was sown with 200 mm row distance to test the tool combinations in well-known conditions. In a second trial, a promising selection of the first one were tested in 150 mm row distance. Both trials were established in winter wheat. Plot size was adjusted to 15 m^2 (1.5 m x 10 m). Weeding was conducted in two to four leaf crop growth stage. As mentioned in chapter 2, parallelograms model "Duo" (K.U.L.T.) were used because of their versatile mounting options. In Figure 3.1 every tool and combination is presented. For all the pre-test trials the manual steered hoe frame model "Argus" (K.U.L.T.) was used.

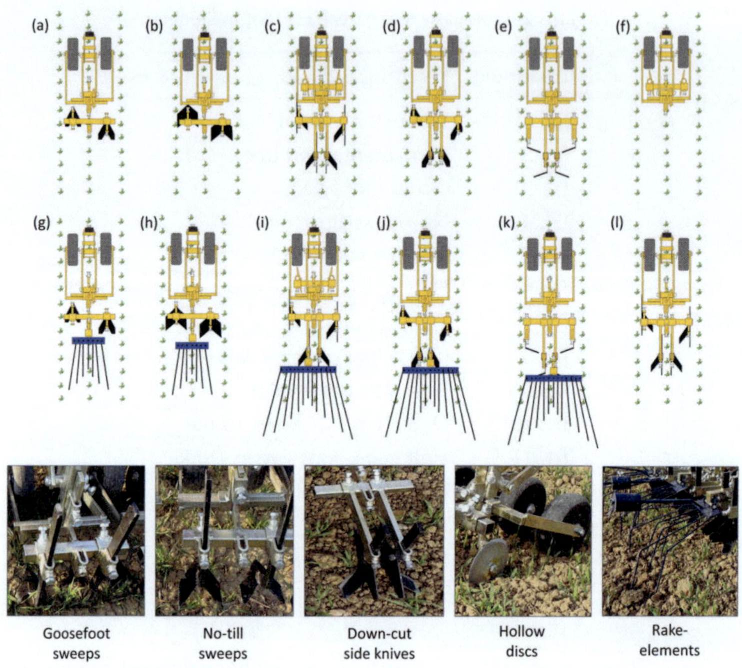

FIGURE 3.1: Overview of the tools for the pre-test. **(a)** Goosefoot sweeps, **(b)** No-till sweeps, **(c)** Down-cut side knifes and hollow discs, **(d)** Half goosefoot sweeps, **(e)** Dam-shape knife, **(f)** Hollow discs, **(g)** Half goosefoot sweeps and rake element, **(h)** No-till sweeps and rake element, **(i)** Down-cut side knifes and rake element, **(j)** Half goosefoot sweeps and rake element, **(k)** Dam-shape knife and rake element **(l)** Down-cut side knifes

(B.Kollenda, K.U.L.T.)

For each treatment, a test run was done outside the experimental plot to proof the alignment and the working depth to 20 mm. The safety distances of the tools to the crop rows were adjusted depending on the tool. Tools cutting along the rows were set to 15 mm distance to the crop, while the goosefoot and the no-till sweeps were set to a distance of 25 mm. Weed density was determined before the tests and right after. For that purpose, weeds were counted in five different 0.1 m^{-2} spots per plot. Weed counting was done separately for inter- and intrarow space. During, mounting and adjustment of the hoe the time consumption was recorded. Also, the homogeneity of the weeding operations per block was determined by a visual scoring after hoeing.

3.4 Details of Experiment 3: Test of selected tools

After the evaluation of tools, a two year trial series was established to confirm the pre-test trials results at various sites and under different weather conditions. In both years, experiments were conducted with a row distances of 200 mm (D1) at HOH, 150 mm (D2) at IHO and 125 mm (D3) at HOH. Selection criteria for row distance were based on the following assumptions. Row spacing of 200 mm in winter wheat represents organic farming practices in Germany. 125 mm and 150 mm are common row distances for the conventional farming systems. Regardless of the row distance, winter wheat planting density was similar in all experiments (350 seeds m^2). Sowing was done with plot equipment described in Chapter 2. The plot size was 1.5 m x 19 m (HOH) and 2 m x 17 m (IHO). An overview of the field trials is given in Table 3.3.

TABLE 3.3: Overview of the row distances, locations, sowing dates and cultivars. **HOH**: Agricultural Experimental Station, Hohenheim. **IHO**: Agricultural Experimental Station, Ihinger Hof. [1]Landbauschule Dottenfelderhof eV, Bad Vilbel, Germany. [2]Getreidezüchtung Peter Kunz, Feldbach, Swiss. [3]Deutsche Saatveredelung, Lippstadt, Germany. [4]R.A.G.T. Saaten, Hiddenhausen, Germany.

Trial ID	Row distance	Site	Cultivar	(Breeder)
D1a	200 mm	HOH	Butaro	(Hartmut Spieß[1])
D1b	200 mm	HOH	Pizza	(Peter Kunz[2])
D2a	150 mm	IHO	Patras	(DSV[3])
D2b	150 mm	IHO	RGT Reform	(RAGT[4])
D3a	125 mm	HOH	Butaro	(Hartmut Spieß[1])
D3b	125 mm	HOH	Rebell	(RAGT[4])

As mechanical weed control devices goosefoot sweeps (GFS) (a), no-till sweeps (NTS) (b) and down cut side knife blades (DSK) (c) (Figure 3.2) were tested in both years. In the 200 mm row distance, the standard tools width of 160 mm was used, while for the smaller distances the blades were modified according to the description in Chapter 2. The DSK blades are designed for pairwise use in each row with the vertical part pulled left and right along the rows. They were developed for hoeing in vegetable crops with wide row spaces. Due to the limited row spaces in these trials, they were only tested in the 200 mm row spaced plots.

FIGURE 3.2: Hoe blades used in the study. **a:** Goosefoot sweep (GFS), **b:** No-till sweep (NTS), **c:** Down-cut side knife (DSK). (B.Kollenda)

Two experiments were conducted in 200 mm row distance, two in a 150 mm row spaced trial and two experiments were conducted with 125 mm row distance. In each experiment, six (150 mm and 125 mm trials) and eight treatments (200 mm trials) were tested Table 3.4. An untreated control without any weeding and a weed-free control achieved through manual weeding or herbicide application were included in all experiments. Hoeing was done with goosefoot sweeps, no-till sweeps and down cut side knife blades at low (3-4 km h^{-1}) and high (6-8 km h^{-1}) speed when winter wheat started tillering and weeds had 2 to 4 true leaves (BBCH 13-21). The 8 km h^{-1} speed was feasible due to the short plot length. In case of major crop damage one of the spare plots were integrated in the trial.

TABLE 3.4: Treatments performed in the experiment. All plots where treated once each season. **GFS**: Goosefoot sweep, **NTS**: No-till sweep, **DSK**: Down-cut side knife; [1]9 g ha^{-1} mesosulfuron + 1.8 g ha^{-1} iodosulfuron (Atlantis WG, 30 g kg-1 mesosulfuron + 6 g kg-1 iodosulfuron, Bayer Crop Science, Germany) [2]48 g ha^{-1} diflufenican + 4.62 g ha^{-1} metsulfuron (Alliance WG 600 g kg^{-1} diflufenican + 57.8 g kg^{-1} metsulfuron , Nufarm, Köln, Germany), [3] 1.8 g ha^{-1} florasulam + 22.5 g ha^{-1} clopyralid (Primus Perfect SC, 25 g l^{-1} florasulam 300 g l^{-1} clopyralid, Bayer Crop Science, Langenfeld, Germany),[4]13.66 g ha^{-1} pyroxsulam 4.56 g ha^{-1} florasulam (Broadway 200 g ha^{-1}68,3 g kg^{-1} pyroxsulam 22,8 g kg^{-1} florasulam Corteva, München, Germany), [5]tribenuron-methyl (Pointer 35 g ha^{-1} (Cheminova, Stade, Germany). Herbicides were applied with a spray volume of 200 l ha^{-1} and 300 kPa pressure. The nozzle type was a 8002 flatfan.

Treatment	Weeding method	2017	2018
Trt. 1	**Untreated**	No treatment	No treatment
Trt. 2	**Weed free**	Manual/Herbicide*	Manual/Herbicide*
Trt. 3	**GFS slow**	3 km h^{-1}	4 km h^{-1}
Trt. 4	**NTS slow**	3 km h^{-1}	4 km h^{-1}
Trt. 5	**DSK slow****	3 km h^{-1}	4 km h^{-1}
Trt. 6	**GFS fast**	8 km h^{-1}	8 km h^{-1}
Trt. 7	**NTS fast**	8 km h^{-1}	8 km h^{-1}
Trt. 8	**DSK fast****	8 km h^{-1}	8 km h^{-1}

*2017: Atlantis[®1], Alliance[®2], Primus[®3]. 2018: Broadway[™4], Pointer[®5].

**DSK were only tested in 200 mm

3.5 Details of Experiment 4

3.5.1 Adjustment of the camera row guidance systems

The hydraulic frame with the camera row guidance system was set up with one parallelogram at the tool bar, equipped with a no-till sweep to determine the accuracy of lateral steering. For each system, different camera adjustments were tested to find the best recognition of the crop rows. Camera position and angle were varied to see whether a wider forward sight or a smaller twist in the field of view leads to better row recognition. When the row recognition was set up a cardboard with the green lines (simulating crop rows) was moved left and right to test the ability of the steering system to follow the artificial rows. At the start of the tests the hydraulic system was set to the lowest possible flow to avoid any overriding of rows. Later tests were done under different hydraulic pressures to investigate the guidance precision under various driving speeds.

3.5.2 Adjusting setup

When the hydraulic steered hoe frame was equipped with one of the two camera systems tests for hoeing accuracy were conducted. Therefore an indoor test bench was set up. The connectors of the three-point hitch-connector were stabilised with two chocks in a way to allow the free movement of the tool bar. A strap secured the top connector of the hitch to a fixation point on the ground. An electric hydraulic pump provided the oil pressure for the steering system. A movable platform was placed under the field of view of the camera. There, a 2x2 m piece of cardboard with green lines was arranged to simulate the crop rows. Light sources were installed on top or at both sides of the camera to simulate different light conditions.

3.5.3 Accuracy test

Both, the 3 m hoe as well as the 6 m segmented hoe were tested for accuracy. For that purpose a tacheometer prism-unit was mounted at the movable part on both models near the camera. The tacheometer was installed in a corner outside of the test field. Maize was used as model crop with a row distance of 750 mm. This was done due to the time of the season and no availability of cereals test area. The measurement frequency of the tacheometer was set to 4 Hz. The hoe was equipped with a common tool set to provide realistic lateral forces on the steering frame. Measurement resolution was around 5 mm. The rows left and right of the plot was also recorded by measuring ten points of every row. Hoeing was performed three times at the same seed-strip to track the deviation of the hoe guidance per run. The analysis was done by calculating the standard deviation to a fitted line.

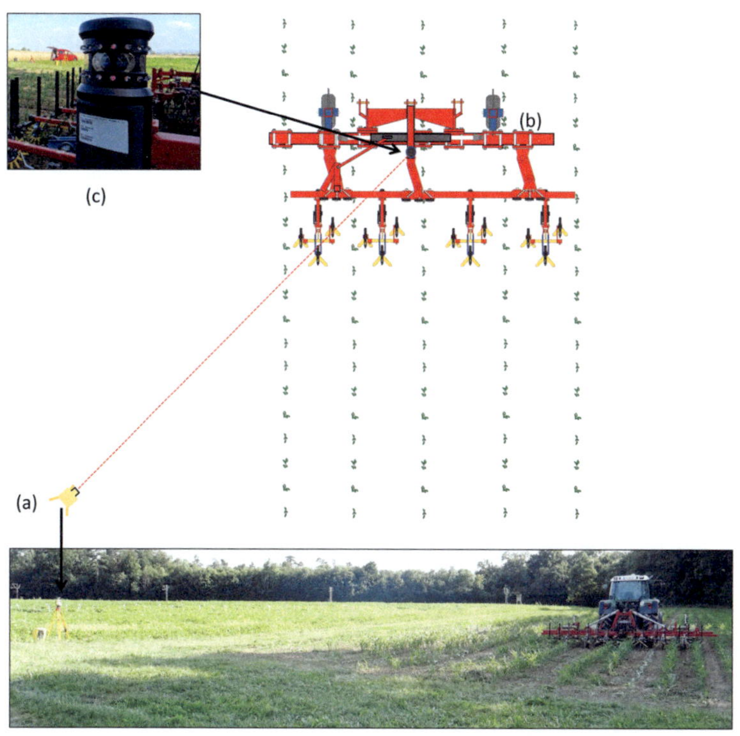

FIGURE 3.3: Tacheometer records of the camera row guidance system. (a) Tacheometer, (b) Camera guided hoe, (c) Prism unit.

3.6 Details of Experiment 5: Test of the prototypes

3.6.1 Test of the 3 m camera guided hoe

Field trials were conducted to test the final set up of the 3 m camera steered prototypes. Winter wheat was sown in October 2018 at 350 seeds m^2. Plot size was adopted to 3 by 30 m with three replicates at HOH and four replicates at IHO. The row distance was 150 mm and sowing was performed by the Deppe

seeder 2 behind a tractor with RTK-GNSS-supported steering to reach straight
row lines.

FIGURE 3.4: 3m camera guided hoe during field trials.

Due to favorable weather conditions first hoeing was conducted already 27th
of February. At IHO the hoeing was done 4th of April. No-till sweeps were
used as hoe tools at 18 mm distance to the center of the crop row on each side.
The working depth of the tools was adjusted to 20 mm and the tool angle was
levelled slightly forward to achieve a better penetration of the relatively crusty
top soil. The treatments were done at three different hoeing speeds and a control
plot with a suitable herbicide application to establish a permanently weed free
control plot as comparison. Another plot was untreated but the same number of
overruns with the tractor as the mechanically treated plots were done to ensure
an even crop damage by the tractor tires (Table 3.5).

TABLE 3.5: Test trial 3 m Hoe Overview of the treatments

Treatments	Weeding method
Trt. 1	Untreated control
Trt. 2	Hand weeding
Trt. 3	Camera Hoe 3 km h^{-1}
Trt. 4	Camera Hoe 6 km h^{-1}
Trt. 5	Camera Hoe 8 km h^{-1}

3.6.2 Test during the night

To test the precision of the camera guidance system at night with artificial light, two light sources (Halogen Spotlites, 12 V, 35 W) were mounted left and right of the camera (Figure 3.5). They were adjusted to shed light in the view of the camera. One was pointed towards driving direction and the other one was mounted to spot 90° to the ground. Software settings, as well as the camera position and the hydraulic adjustments, were the same as during daylight.

FIGURE 3.5: Setup for night tests with two halogen spotlights.

3.6.3 Test of the 6 m segmented hoe

Summer barley was sown at with 400 seeds m^2 in a row distance of 150 mm in 3 m strips. Sowing was done by RTK-GNSS-steered equipment linked to a trial map that pretends a pre-adjusted 100 mm row between two beds. This was done to test the ability of the anti-collision system in a worst case scenario. Two 3 m sowing strips were combined to a 6 m wide test plot for the 6 m segmented hoe. The plot length was 80 m.

TABLE 3.6: Test trial 6 m hoe. Overview of the treatments

Treatments	Weeding method
Trt. 1	Untreated control
Trt. 2	6 m Hoe 3 km h^{-1}
Trt. 3	6 m Hoe 10 km h^{-1}

FIGURE 3.6: 6m segmented hoe in a field trial.

3.7 Details of Experiment 6: Integrated weed management trials

Once successful weed control with the hoe was confirmed further trials were initiated to test the fit of the new hoe as a part for integrated weed management approaches. Therefore, hoeing was combined with a harrowing as well as with a herbicide applications. The herbicide application was used to control weeds that are typically problematic by the ability to climb up the crop and being not efficient suppressed by the closing canopy. In the trials before this was mainly *P. convolvulus* and *G. aparine*. For the weed-free control plot, a suitable herbicide mixture was used while in the hoe and herbicide combination just the active ingredients against these problematic weeds where sprayed.

A trial series was conducted for two years. In each year two sites were sown with winter wheat with a row distance of 150 mm. In 2018 the two sites were located at IHO. In 2019 one site was located at IHO while the other site was at HOH. Sowing rate was 370 seeds m^2. Seed bed preparation was done by two times flat tilling with a cultivator and just before seeding two times treated by a rotary harrow. Treatments in 2018 where done in late April while in 2019 it was performed in late February at HOH and mid-March at IHO. An overview of the treatments is presented in Table 3.7.

TABLE 3.7: Overview of the weed management trials. [1]9 g ha^{-1} mesosulfuron + 1.8 g ha^{-1} iodosulfuron (Atlantis® WG, 30 g kg^{-1} mesosulfuron + 6 g kg^{-1} iodosulfuron, Bayer Crop Science, Germany), 4 g ha^{-1} metsulfuron + 16 g ha^{-1} carfentrazon (Artus® WG, 100 g kg^{-1} metsulfuron + 400 g kg^{-1} carfentrazon, Cheminova, Germany), 3 g ha^{-1}(Saracen® SC 50 g L^{-1}, Nufarm, Germany). [2]0.037 L ha^{-1} florasulam, 0.15 L ha^{-1} fluroxypyr (StaraneXL® SC 2.5 g L^{-1} florasulam, 100 g L^{-1} fluroxypyr, Dow AgroScience, Germany). [3]136 g ha^{-1} pyroxsulam, 45.6 g ha^{-1} florasulam (Broadway WG, 68.3 g kg^{-1} pyroxsulam, 22.8 g kg^{-1} florasulam, Dow AgroScience, Germany), 17.5 g ha^{-1} tribenuron (Pointer® WG, 500 g kg^{-1} tribenuron, Cheminova, Germany). [4] 136 g ha^{-1} pyroxsulam, 45.6 g ha^{-1} florasulam (Broadway WG, 68.3 g kg^{-1} pyroxsulam, 22.8 g kg^{-1} florasulam, Dow AgroScience, Germany). Herbicides were applied in 200 l ha^{-1} and 300 kPa pressure. The nozzle type was a 8002 flatfan

Year	Treatment	Strategy
	Trt. 1	Untreated control
	Trt. 2	Herbicide[1]
2018	Trt. 3	Hoe (6 km h^{-1})
	Trt. 4	Harrow (9 km h^{-1})
	Trt. 5	Hoe (6 km h^{-1}) + Herbicide[2]
	Trt. 6	[1]Atlantis®, Artus®, Saracen® [2]StaraneXl®
	Trt. 1	Untreated control
	Trt. 2	Herbicide[3]
2019	Trt. 3	Harrow (9 km h^{-1})
	Trt. 4	Hoe (8 km h^{-1})
	Trt. 5	Hoe (8 km h^{-1}) + Harrow (9 km h^{-1})
	Trt. 6	Hoe (8 km h^{-1}) + Herbicide[4]
		[3]Broadway™, Pointer® [4]Broadway™

When weeding was conducted the development stage of the winter wheat time was always around three to four leave stage. Treatments were done only one time per season. In 2018 the hoe treatments were done by a manual steered 3 m Argus hoe and in 2019 the treatments were conducted with the 3 m camera steered prototype. Therefore, driving speed of hoeing was 6 km h^{-1} in 2018 and 8 km h^{-1} in 2019 with the camera guided hoe. Harrowing was done with a driving speed of 9 km h^{-1}. In 2018 the trials were finished after the treatment effect was recorded, while in 2019 also the grain yield was determined.

Chapter 4

Results and Discussion of the Experiments

4.1   Experiment 1: Crop soil cover measurements

After the adjustments covered the tested set of images, the crop soil cover (CSC) measurements were used to determine the amount of soil that buried the crop plants for each tool. As shown in Figure 4.1, various cameras can be used to properly separate between plant material and the background. The results were stable irrespective of the light conditions, row distance and within a wide crop growth range. The algorithm used for that purpose, was discriminating crop and soil without additional modifications. Furthermore, small weeds in the inter-row space are successfully deleted by this method and did not influence the crop cover analysis. The discrimination of the algorithm between crop and soil was stable over multiple images without modifications, as long as the plants were not densely developed and the leaf cover was below 70 %. The maximum deviation between hand segmented and algorithm processed images was 0.69 % for the compact camera image set and 2.84 % for the smartphone camera image set. A certain deviation was noticed between the algorithm estimations of the hand-held camera and the smart phone. The algorithm constantly underestimated the leaf cover of images captured with the smart phone camera. In consequence, this affected the CSC-calculation, but the deviation was constant and could be correlated to the sensor. The present methodology is fast and easily affordable in terms of equipment and resources. Therefore, it is a viable replacement for previously used methods, where either wooden sticks were used to measure the amount of soil burying (Zhang and Chen, 2017) or by visual scoring.

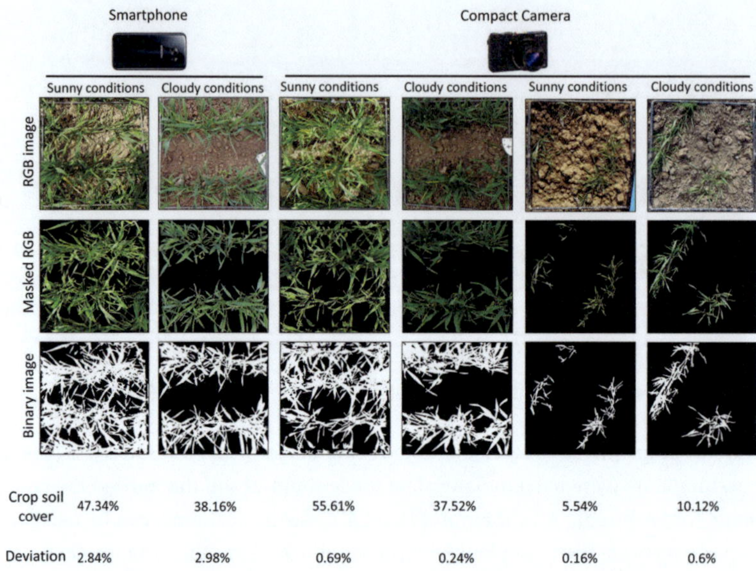

FIGURE 4.1: Crop soil cover results of some test images. The two images on the left were captured by a smartphone model SM-G930F (Samsung Electronics Co., Suwon, South Korea). The rest of the images are taken by a DSC-HX60V compact camera (Sony Corp., Minato, Tokyo, Japan). The top row shows the original RGB image. The middle row shows the masked image where just plant material is visible. At the bottom the binary images are presented. In the next row under the images the crop soil cover calculated by the algorithm is given and in the last row the deviation to a manual segmentation is presented. Source of the images of the smartphone: `https://www.samsung.com/de/smartphones/galaxy-s7/overview/`. Source of the image of the Camera: `https://www.sony.de/electronics/cyber-shot-kompakt-kameras/dsc-hx60-hx60v`.

4.1.1 Discussion of Experiment 1

According to Rasmussen (1997), the variation of crop soil cover estimations by visual scoring can be 20 % even if technicians are trained. Counting the crop density decrease after mechanical weeding is not possible due to the fact that the crop is covered with soil. Therefore, Jensen *et al.* (2004) proposed crop soil cover determination as the best alternative. Instead of using a threshold-based separation algorithm (Rasmussen *et al.*, 2007), the segmentation approach used in this study was stable across a wide range of light conditions and plant densities. Thus, any number of sample pictures is processable as the calculation is done automatically and last only a few seconds per image. To use the manual weed control as an uncovered reference value, proofed to be a proper solution. If hand weeding was performed, the ripped off weeds were completely removed from the site and in cases a herbicide application was used, images were taken when weeds were already dried down. Those dead weeds were recognised as background by the segmentation algorithm and did not interfere with the plant density calculations. Furthermore, the crop soil cover is a value that showed a real influence on the crop and is more accurate compared to a human eye crop cover estimation. But the proposed approach depends on the crop growth stage and the resolution of the used sensors. As long as the plant material and the ratio between plant pixels and background pixels is within a certain range, the algorithm provides reliable results. Images recorded with other sensors, from various perspectives or distances above ground, can be also processed after a software re-calibration. One of the main advantages of the new method is that the complete algorithm can be integrated in a portable smart device, that can estimate the crop soil cover by combining the technical specification of the sensor system and the distance to the ground. When compared to studies of other authors, the proposed algorithm performed at the same level of accuracy. For example the method proposed by Bai *et al.* (2013) which was used to discriminate between rice-plants and background. Even though they used another colour space (LAB), the discrimination accuracy was on the same level. In contrast to that, Ruiz-Ruiz *et al.* (2009) used a segmentation developed to classify plant species where more background noise remained. Similar problems are encountered by Tian and Slaughter (1998). Even though their recognition database is much larger and diverse, as well as providing a more robust result than the image database used in this study, the segmentation quality seemed to be lower. The new method described in this study also differs from the highly precise segmentation proposed by Onyango and Marchant (2003) where small weeds remained visible after processing. To meet the goal of the new method to determine the leaf cover of winter wheat and to calculate the crop soil cover, it was important to delete the weeds from the images. However, when used in an early growth stage can attribute in error, small crop plants as weeds. As this was steady, it was valued as not important and the CSC-measurement was used in every hoeing test to determine the selectivity and the influence on the crop. During the experiments dealing with the differences between hoe tools, the algorithm was a support. Using the compact camera the calculated CSC was within a measurement accuracy of less than +/- 1 %. It provided important results that were used in the decision-making process of the following Experiments to find the best tool.

4.2   Experiment 2: Pre-test of tools

The results of the first pre-test of hoe tools showed that some tool combination were usable in cereal crops. Others had major limitations even after their modifications. In the following chapter all weed control efficacy results are described. Figure 4.2 shows the results for the single tools.

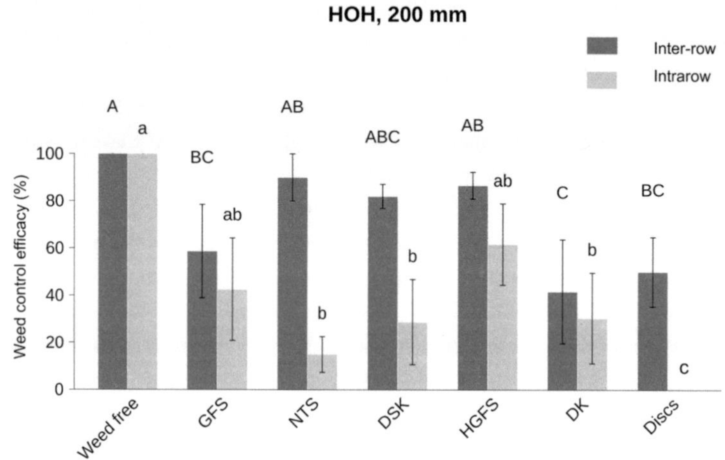

FIGURE 4.2: Weed control efficacy results of the single tools in 200 mm row width at Hohenheim. The dark grey bars represent the efficacy between the crop rows while the light grey bars show the intrarow weed control efficacy. Means with the same letters are not significantly different according to Duncan's multiple range test at $\alpha \leq 0.05$. Capital letters represent significance of the inter-row space and small letters show the values between the cop rows. The standard error of the mean (SEM) is represented by the small black bars.

In the inter-row space, goosefoot sweeps and discs resulted in a significantly smaller control success with an average of 60 % compared to the weed free control plots which were hand weeded. The other tools were at the same level with more than 80 % weed control efficacy. Only the no-till sweeps reached in some plots comparable levels of control success then the weed free control plot. In the intrarow space, the no-till sweeps, half-goosefoot sweeps and the dam-shape knifes did not reach the level as in weeding as the manual weeded control plots. The discs had no effect on the intrarow weed control efficacy.

In some plots, the rake elements enhanced the inter-row weed control efficacy of the goosefoot sweeps up to 100 %. The no-till sweeps in combination with the rake elements lowered the weed control level to 65 % on average. All inter-row results are not significantly different to the manual weeded plots. In the intra row space, the rake elements did not lead to a significant better weeding. For further details see Figure 4.3.

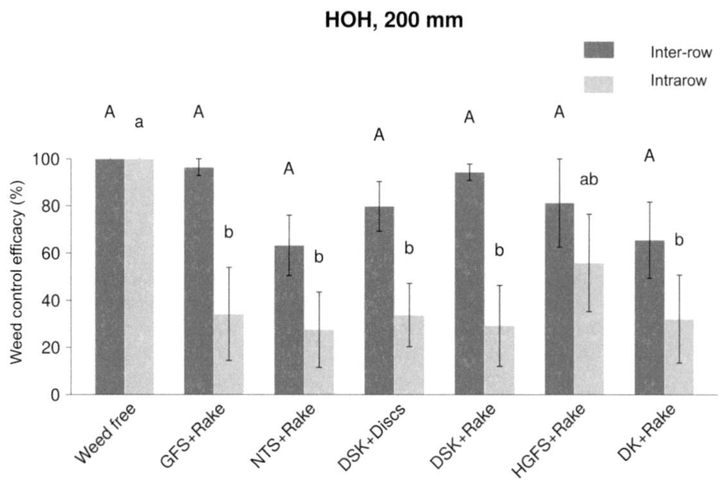

FIGURE 4.3: Weed control efficacy results of the test trials in 200 mm row width at Hohenheim (HOH). The dark grey bars represent the efficacy between the crop rows while the light grey ones show the intrarow weed control efficacy. Means with the same letters are not significantly different according to Duncan's multiple range test at $\alpha \leq 0.05$. Capital letters represent significance of the inter-row space and small letters show the values between the cop rows. The standard error of the mean (SEM) is represented by the small black bars.

The test of some promising tools in 200 mm row width showed similar results in 150 mm row distance. Except the combination of no-till sweeps and rake, all tools performed at the same level as the manual weed control in the inter-row space. For the intrarow space, the weed control success was at similar levels whether with the rake elements or without. Only the no-till sweeps performed again better without the rake-elements (Figure 4.4).

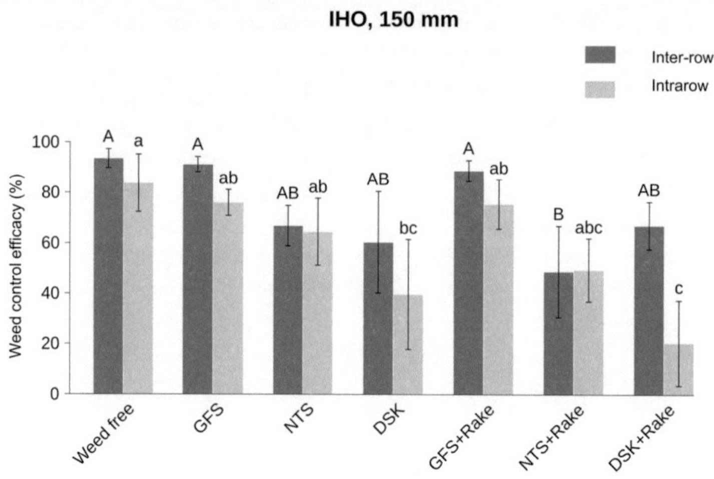

FIGURE 4.4: Weed control efficacy results of the test trials in 150 mm row width at Ihinher Hof (IHO). The dark grey bars represent the efficacy between the crop rows while the light grey ones show the intrarow weed control efficacy. Means with the same letters are not significantly different according to Duncan's multiple range test at $\alpha \leq 0.05$. Capital letters represent significance of the inter-row space and small letters show the values between the cop rows. The standard error of the mean (SEM) is represented by the small black bars.

Concerning handling, the goosefoot sweeps and the no-till sweeps positioned in the center of the inter-row space performed better in terms of adjustment time. The goosefoot sweeps delivered the largest amount of soil movement from inter to intrarow space. Therefore, the passive weeding effect by burying weeds with soil can be considered very efficient when using goosefoot sweeps. On the other hand, the effective working width can be broader than the size of the tool, which may cause crop damage when operated with higher driving speeds, especially in crusty soil areas. On that point, the no-till sweeps are better in a wider range of driving speeds. They also seemed to perform mostly identical across all replications compared to the other tools. Especially the goosefoot sweeps showed different amounts of crop soil cover. Any other tool setup including the discs was a lot more complicate to adjust. Another issue was that the pairwise mounting of main tools asks for a set up where the tools required their arrangement behind each other due to the limited space between the rows. The distance between the tools was important to allow any biomass residues, ripped out weeds and soil to flow through the tools to prevent clogging of the hoe tool unit. The larger distance from the pivot point of the parallelogram to the furthest resulted in a widely varying working depth even after time consuming adjustments. Reduced precision was identified across the different blocks under various soil conditions and required additional adjustments. In cases of problematic soil conditions and small crop plants, the down-cut side knife showed good crop selectivity and an increased CSC cutting the topsoil at the boarder to the intrarow space and by leaving an untouched soil strip around the crop. The dam-shape knifes performed better in handling compared to the down-cut side knifes, but the material of the blade was too thin and thus prone to bending. The rake elements behind the main tools performed well concerning handling and adjustments. In light soil and low operating speeds, the rake element might increase the weed control efficacy, but with this specific setup it did not penetrate hard soil patches which can occur at sites with heavy soil. Furthermore, the rake elements started jumping when operating speed was faster than 5 km h^1.

4.2.1   Discussion of Experiment 2

To screen various available tools and shape of tools and modify them for the use in narrow seeded cereals, was expected to be faster and more efficient compared to develop new tools. The hollow-discs did not enhance weeding performance significantly. Nevertheless, when using direct seeding techniques, the discs might become useful to prevent clogging due to biomass residues on the soil. The weed control efficacy in the intrarow space depends on the amount of soil that is moved into the crop row. This was also described in other studies (Van Der Weide *et al.*, 2008; Zhang and Chen, 2017). The goosefoot sweep moved more soil in the intrarow space as the no-till sweep. These findings are in line with the investigations of Pullen and Cowell (1997); Zhang and Chen (2017). But

Zhang and Chen (2017) also described a large burying effect by the use of partially cut goosefoot sweeps. This was not confirmed in our study. The weeding performance of harrow equipment is well described in the literature (Rasmussen, 1991; Pullen and Cowell, 1997; Lötjönen and Mikkola, 2000; Melander *et al.*, 2003; Jacobsen *et al.*, 2010) and was expected to be a sufficient way to complement and to increase the weed control efficacy of hoeing. The disappointing results of the rake element are most probably due to their light design. Therefore, the tension of the tines was not strong enough to further weed through the soil in the intrarow space, or crush bigger soil chunks with well-rooted weeds inside. Additionally, it was observed that the rake elements moved soil out of the intrarow space, which uncovered already buried weeds. As they were designed for the use in vegetable cropping where soil conditions are more loose compared to cereal cropping the rake elements could not perform sufficiently. These results might be linked to some other insufficient weed control efficacy results for tools that were not shaped to move soil into the crop row (e.g. Half-goosefoot sweeps). The selection was driven either by their efficiency to move soil in the crop row while other models are specifically tested on their ability to save the crop row from too much soil cover. This was done to find an improved hoe tool that performs well over a wide range of conditions. Another aspect was the adjustment performance and the versatility of a tool to reach the best possible weed control. Therefore, the goosefoot and the no-till sweeps selected as the most promising tools after that test. Based on these results, a decision was made for the consecutive trials: The goosefoot and the not-till sweeps were selected for the next tests with various driving speeds to understand in which driving speed which tools perform best. Beside that, down-cut side knifes were also tested further because of the effort saving the crop row in early crop growth stages and under humid or crusty soil conditions. All other tools will not be used in the next trials to put the focus on the most promising ones and to reduce the amount of treatments to allow more sophisticated analysis.

4.3 Experiment 3: Test of selected tools

4.3.1 Weed species composition

In the six field trials, mainly broad-leaved weed species occurred. In all trials, the average weed infestation was typical for fields of conventional farming except that there were no major grass weed species. An overview of the three most occurred weed species per trial is given in Table 4.1.

TABLE 4.1: Weed species composition and densities in the trials. Trial ID's are named according to Table 3.3. **D1a**: 200 mm, Hohenheim, 2017.**D1b**: 200 mm, Hohenheim, 2018. **D2a**: 150 mm, Ihinger Hof, 2017. **D2b**: 150 mm, Ihinger Hof, 2018. **D3a**: 125 mm, Hohenheim, 2017. **D3b**: 125 mm, Hohenheim 2018

Trial ID	Weed Species	Density (m^{-2})
D1a	*S. arvensis* L.	14.4
	S. media L.	7.5
	P. convolvulus L.	4.5
D1b	*M. chamomilla* L.	75.1
	S. media L.	33.4
	G. aparine L.	12.1
D2a	*P. convolvulus* L.	6.2
	G. aparine L.	4.2
	S. arvensis L.	1.6
D2b	*V. persica* L.	94.6
	M. chamomilla L.	16.8
	L. purpureum L.	8.4
D3a	*S. arvensis* L.	14.4
	S. media L.	7.5
	P. convolvulus L.	7.4
D3b	*L. purpureum* L.	28.9
	C. hirsuta L.	10.9
	V. persica L.	10.2

At the Hohenheim site in 200 mm row distance, only small amounts of weeds had emerged. At trial D1a *Sinapis arvensis* L. (field mustard) occurred as the most abundant species with 14 plants m^{-2}. In 2018 trials weed densities were higher than in the previous year. At D1b *Matricaria chamomilla* L. (scented mayweed) was recorded with 75 plants m^{-2}. At the Ihinger Hof site with 150 mm row distance, the weed pressure was low in 2017. In D2a the most frequent species with an average density of 6.2 plant m^{-2} was *P. convolvulus..* In 2018, at D2b weed densities were higher than in the previous year. *Veronica persica* (common field speedwell) appeared with an average of 94.6 plants m^{-2}. D3a (125 mm row distance) was conducted in the same field in Hohenheim as D1a and weed density was similar. At trial D3b the most abundant species was *Lamium purpureum* L. (red dead-nettle) with 28.9 plants per m^{-2} before treatment.

4.3.2 Weed control efficacy

Between the crop rows, the weed control efficacy of all hoe treatments compared to the untreated control plots lead to a weed reduction of 65 % to 98 %. The separately assessed intrarow weed control efficacy varied from 24 % to 80 %. For both intra- and inter-row space, the hand weeded plots showed a weed control efficacy of 97 % in 200 mm and around 90 % in 125 mm row width trials. In 150 mm row width where the herbicide was applied, a weed control performance of 96 % between the rows and 82 % in the row was measured.
In 200 mm row distance trails fast driven GFS and DSK lead to an average inter-row weed control efficacy comparable to the hand weeded control plots (Figure 4.5). The weed control efficacy of the other hoe treatments was significantly lower in the crop rows compared to the weed free control. The most effective weed control was performed by the fast GFS treatments. In the intrarow space, the weed control efficacy of all hoe tools was at the same level significantly lower than the hand weeding except the slow driven NTS which was significantly lower compared to all other treatments.

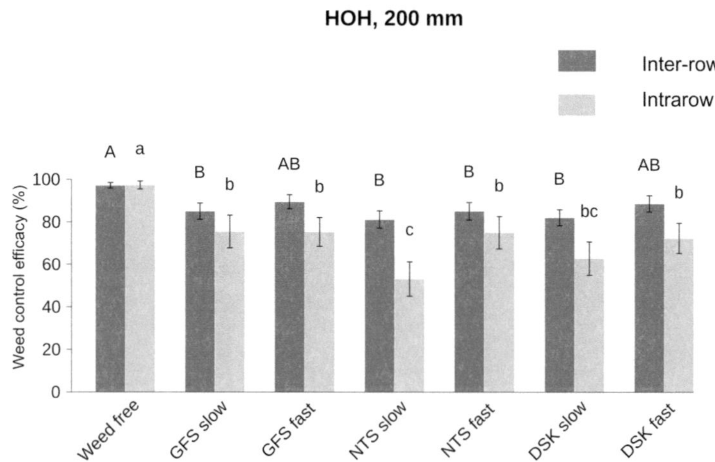

FIGURE 4.5: Weed control efficacy of the tool trials in 200 mm row width. Data is pooled from both trial years. The dark grey bars represent the efficacy between the crop rows while the light grey bars show the intrarow weed control efficacy. **Weed free:** Hand weeding, **GFS:** Goosefoot sweep, **NTS**: No-till sweep, **DSK**: Down-cut side knife. Means with the same letters are not significantly different according to Duncan's multiple range test at $\alpha \leq 0.05$. Capital letters represent significance of the inter-row space and small letters show the values between the cop rows. The standard error of the mean (SEM) is represented by the small black bars.

Also in 150 mm the GFS showed the best weed control efficacy of 97 % (slow) and 98 % (fast) in the inter-row area at both velocities. Similar results of 94 % weed control efficacy were achieved where slow hoeing was performed with NTS (Figure 4.6). The same occurred in the plots were the hoe equipped with NTS was operated at higher velocity. In general weed control efficacy in the intrarow were not significantly different from each other. They varied between 70 % to 82 % on average. Only NTS fast was significantly lower (60 %).

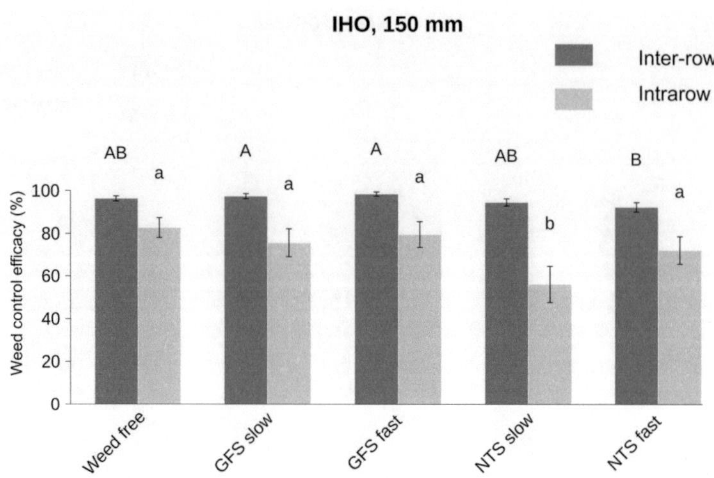

FIGURE 4.6: Weed control efficacy of the tool trials in 150 mm row
width. Data is pooled from both trial years. The dark grey bars
represent the efficacy between the crop rows while the light grey
bars show the intrarow weed control efficacy. **Weed free:** Hand
weeding, **GFS:** Goosefoot sweep, **NTS:** No-till sweep, **DSK:** Down-
cut side knife. Means with the same letters are not significantly
different according to Duncan's multiple range test at $\alpha \leq 0.05$.
Capital letters represent significance of the inter-row space and
small letters show the values between the cop rows. The standard
error of the mean (SEM) is represented by the small black bars.

In trials seeded with 125 mm row distance the inter-row weed control efficacy
showed wider variation. The GFS fast was again on the same significance level
as the manual weeded control at 90 %. GFS slow and NTS fast were significantly
lower (80 %) than the weed free control and NTS slow was significantly lower
compared to the other hoe treatments with just 60 % control efficacy on average.

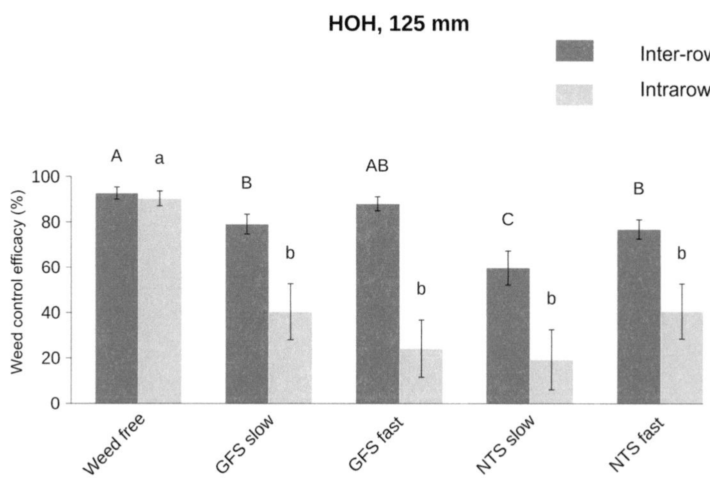

FIGURE 4.7: Weed control efficacy of the tool trials in 125 mm row width. Data is pooled from both trial years. The dark grey bars represent the efficacy between the crop rows while the light grey bars show the intrarow weed control efficacy. **Weed free:** Hand weeding, **GFS:** Goosefoot sweep, **NTS:** No-till sweep, **DSK:** Down-cut side knife. Means with the same letters are not significantly different according to Duncan's multiple range test at $\alpha \leq 0.05$. Capital letters represent significance of the inter-row space and small letters show the values between the cop rows. The standard error of the mean (SEM) is represented by the small black bars.

4.3.3 New emergence of weeds after hoeing

The assessment of weeds relative to the manual weeded control that survived treatments and newly emerged ones two weeks after the treatments showed differences across the row distances (Figure 4.8, 4.9, 4.10). In 200 mm, the untreated control appeared balanced with 27 new emerged weeds m^{-2}) and 25 weeds m^{-2} counted as further developed and therefore remained from the scoring directly after treatments. More survivors (15 weeds m^{-2}) than newly emerged weeds (3-7 weeds m^{-2}) were found in the hoed plots. Hoeing with DSK and NTS seemed to stimulate more weeds to emerge compared to the GFS (Figure 4.8).

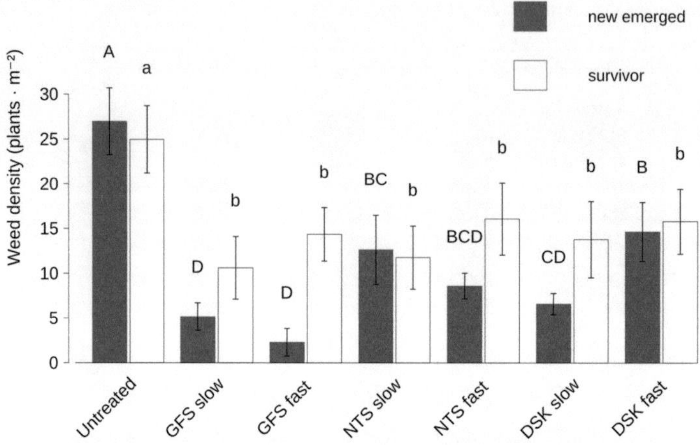

FIGURE 4.8: Composition of new emerged weeds and surviving weeds in 200 mm row width 14 days after treatments relative to the manual/herbicide control. Data is pooled from both trial years. **Untreated:** No treatment, **GFS:** Goosefoot sweep, **NTS:** No-till sweep, **DSK:** Down-cut side knife. Means with the same letter are not significantly different according to Duncan's multiple range test at $\alpha \leq 0.05$. Capital letters represent significance of new emerged weeds and small letters show the level for weeds that survived. The standard error of the mean (SEM) is represented by the small black bars.

In 150 mm row distance less newly emerged (7–10 weeds m^{-2}) were recorded in the hoed plots (except for GFS slow) 14 days after treatments. In the untreated control the amount of new emerged weeds was recorded with 16 plants m^{-2} and 18 weeds m^{-2} were remained. on average the plots treated with NTS showed more newly emerged weeds compare to the GFS treated plots. Further on, the fast-operated treatment always lead to more new emerged weeds on average. The GFS slow treatment show significantly less new emergence of weeds compare to the other hoe treatment. Survivors were at the same level of significance over the all hoe treated plots (Figure 4.9).

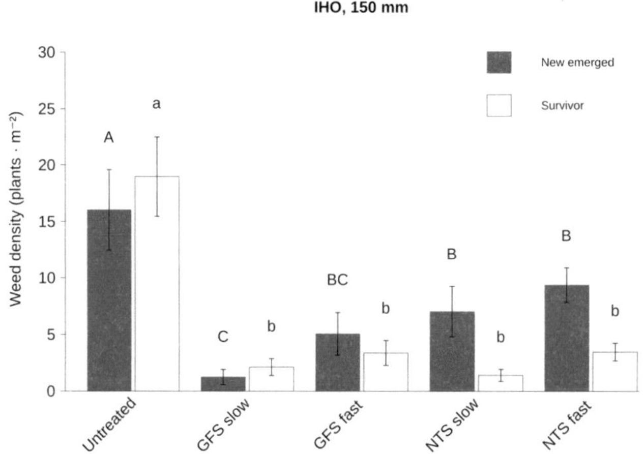

FIGURE 4.9: Composition of new emerged weeds and surviving weeds in 150 mm row width 14 days after treatments relative to the manual/herbicide control. Data is pooled from both trial years. **Untreated:** No treatment, **GFS:** Goosefoot sweep, **NTS:** No-till sweep, **DSK:** Down-cut side knife. Means with the same letter are not significantly different according to Duncan's multiple range test at $\alpha \leq 0.05$. Capital letters represent significance of new emerged weeds and small letters show the level for weeds that survived. The standard error of the mean (SEM) is represented by the small black bars.

Fourteen days after treatment the lower weed control efficacy in 125 mm row distance was still observable. Many more survivors (40–50 weeds m^{-2}) occurred after hoeing and just a few (less than 10 weeds m^{-2}) new weeds emerged. In the untreated control plots an average of 10 new emerged weeds m^{-2} and 55 weeds m^{-2} further developed ones were counted. The NTS slow treatment lead to significant lower new emerged weeds than the GFS slow treatments (Figure 4.10.

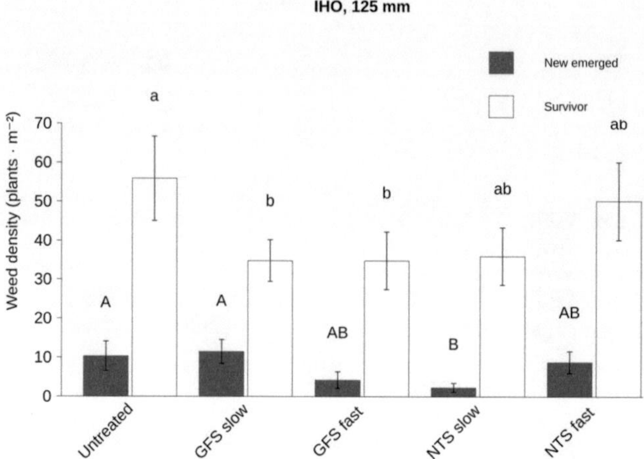

FIGURE 4.10: Composition of new emerged weeds and surviving weeds in 125 mm row width14 days after treatments relative to the manual/herbicide control. Data is pooled from both trial years. **Untreated:** No treatment, **GFS:** Goosefoot sweep, **NTS**: No-till sweep, **DSK**: Down-cut side knife. Means with the same letter are not significantly different according to Duncan's multiple range test at $\alpha \leq 0.05$. Capital letters represent significance of new emerged weeds and small letters show the level for weeds that survived. The standard error of the mean (SEM) is represented by the small black bars.

4.3.4 Selectivity and Crop Soil Cover

Covering weeds with soil in the intrarow space was a major goal of inter-row hoeing. But this may also negatively impact crop growth and yield. The absence of this negative impact together with at the same time good weed control was defined as selectivity. The results on selectivity are presented only for trials with 200 mm and 150 mm row width (Table 4.2). In 125 mm row distance the earlier crop canopy closure did not allow to measure CSC. The most selective weed control in all trials was performed by NTS except in 2017 where GFS fast was slightly better. There was a trend that slow driven tools lead to higher selectivity values. The DSK which were used in 200 mm were less selective than the other tools.

TABLE 4.2: Selectivity as the ratio of weed control efficacy and the amount of crop soil cover (CSC). Results are shown for hoe blades in 200 mm and 150 mm row distance trials. Results with the same superscript letters are not significantly different according to Duncan's multiple range test at $\alpha \leq 0.05$.

	2017		2018	
Treatment	**200 mm**	**150 mm**	**200 mm**	**150 mm**
GFS slow	2.85^b	2.86^b	7.82^a	5.04^a
GFS fast	7.94^s	4.21^b	2.82^a	6.1^b
NTS slow	6.9^a	8.27^a	11.96^a	12.35^{ab}
NTS fast	3.0^b	4.3^b	4.34^a	15.8^a
DSK slow	2.73^b	-	1.36^a	-
DSK fast	2.42^b	-	1.66^a	-

In 200 mm row distance the fast-operated hoe treatments mostly lead to a higher crop cover (Figure 4.11). In 2017 CSC was higher than in 2018 for all treatments. During the first trial period the highest CSC rate resulted in plots treated with the GFS fast while in 2018 the DSK fast brought more soil movement into the intrarow space. The lowest burying with soil was recorded by the use of NTS slow in both years.

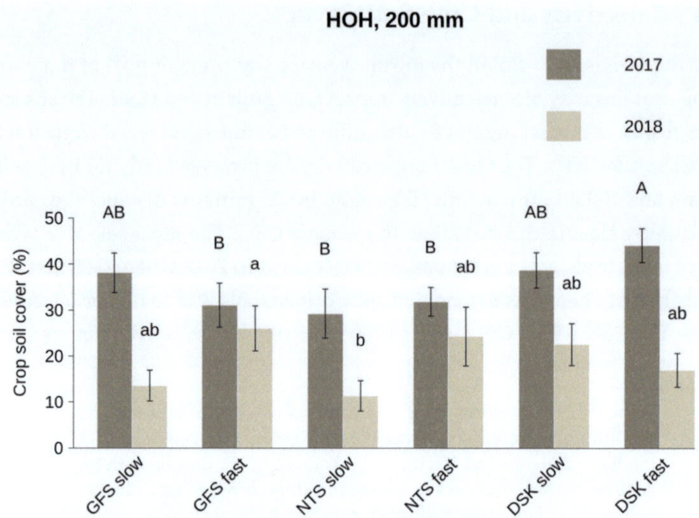

FIGURE 4.11: Crop soil cover (CSC) of the tool trials in 200 mm row width. The dark grey bars represent the CSC 2017 while the light bars show the amount of ridging in 2018. **GFS:** Goosefoot sweep, **NTS**: No-till sweep, **DSK**: Down-cut side knife. Means with the same letter are not significantly different according to Duncan's multiple range test at $\alpha \leq 0.05$. Capital letters represent significance of the trial year 2017 and small letters show the values of 2018. The standard error of the mean (SEM) is represented by the small black bars.

At 150 mm in the year 2017 there was more soil moved in the crop row (20-40 %) while GFS showed more CSC than NTS in both trial seasons. But the slow versions always caused more ridging than the fast versions (Figure 4.12).

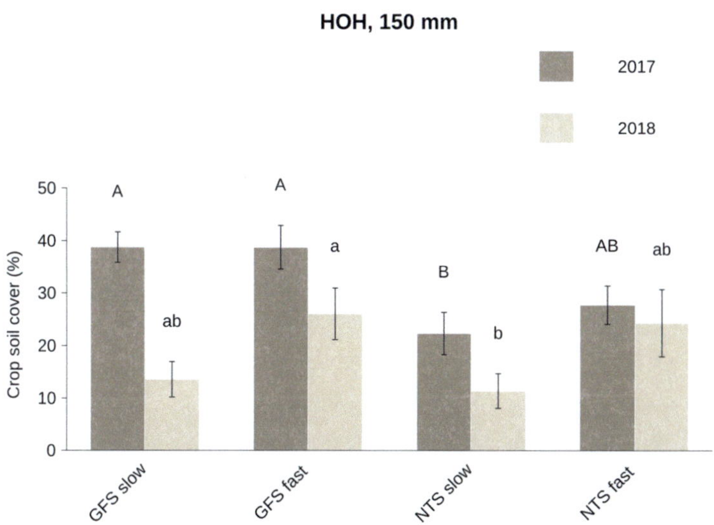

FIGURE 4.12: Crop soil cover (CSC) of the tool trials in 150 mm row width. The dark grey bars represent the CSC 2017 while the light bars show the amount of ridging in 2018. **GFS:** Goosefoot sweep, **NTS:** No-till sweep, **DSK:** Down-cut side knife. Means with the same letter are not significantly different according to Duncan's multiple range test at $\alpha \leq 0.05$. Capital letters represent significance of the trial year 2017 and small letters show the values of 2018. The standard error of the mean (SEM) is represented by the small black bars.

4.3.5 Grain yield

Similar grain yields were observed in trials with the same row distance. Slightly lower yields were recorded for the mechanically treated plots. However, losses were rarely significant compared to the permanently weed free control. The untreated plots had the lowest grain yield in 150 mm and 125 mm row distance. The yield decrease with a difference of minimum 2 t ha^{-1} in the wider row space of 200 mm was confirmed in the result of this study. In 200 mm row distance trials, the average yield of the GFS treated plots was lower than in the untreated control plots. In 2017, the highest yield was measured in the weed free plots. However, hoeing with NTS showed even higher yields in plots planted with 200 and 125 mm row distance in 2018 (Figure 4.13).

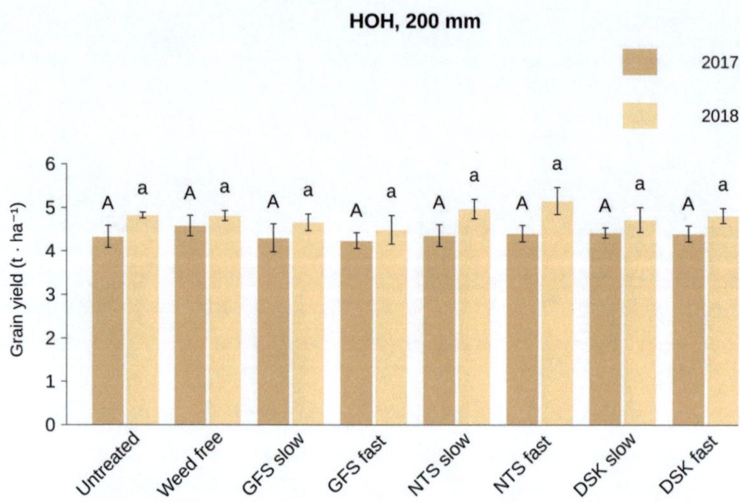

FIGURE 4.13: Grain yield of the tool trials in 200 mm row width. The dark yellow bars represent the yield of 2017 while the light yellow bars show the amount of harvest in 2018. **Untreated:** No treatment, **Weed free:** Manual weeding, **GFS:** Goosefoot sweep, **NTS:** No-till sweep, **DSK:** Down-cut side knife. Bars with the same letters. Means with the same letter are not significantly different according to Duncan's multiple range test at $\alpha \leq 0.05$. Capital letters represent significance of the trial year 2017 and small letters show the values of 2018. The standard error of the mean (SEM) is represented by the small black bars.

Also in 2018, within the same row distance, yield of the mechanical weed control plots did not differ significantly. But in both years the lowest yields (4.4 t ha^{-1}, 2017; 5 t ha^{-1}, 2018) were recorded in the 200 mm spaced trials followed by the 150 mm seed distance with 8 t ha^{-1} (Figure 4.14).

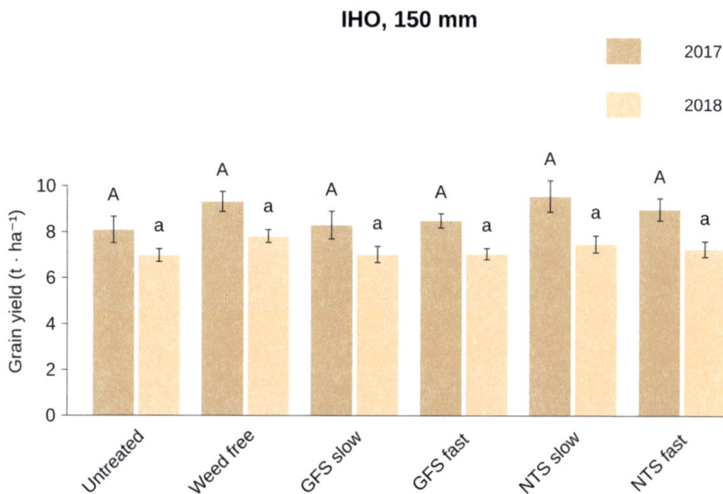

FIGURE 4.14: Grain yield of the tool trials in 150 mm row width. The dark yellow bars represent the yield of 2017 while the light yellow bars show the amount of harvest in 2018. **Untreated:** No treatment, **Weed free:** Manual weeding, **GFS:** Goosefoot sweep, **NTS:** No-till sweep, **DSK:** Down-cut side knife. Bars with the same letters. Means with the same letter are not significantly different according to Duncan's multiple range test at $\alpha \leq 0.05$. Capital letters represent significance of the trial year 2017 and small letters show the values of 2018. The standard error of the mean (SEM) is represented by the small black bars.

Over all, the highest yields were recorded in the 125 mm row spacing (up to 10 t ha⁻¹). Yields in untreated plots showed a decrease from narrow to wide row spaces. Additionally, at 150 mm and 125 mm row width, there is a trend that shows that mechanical weed treatments with a higher speed performed better compared to treatments with the slower velocity. If weeding was performed with the NTS fast, the plots showed the highest yield across all mechanical treatments. With 125 mm row distance significant higher yield was recorded in these plots compared to the untreated control (Figure 4.15).

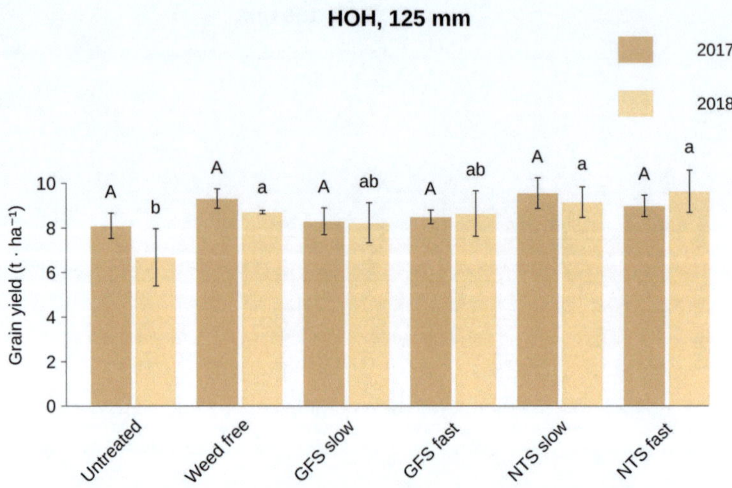

FIGURE 4.15: Grain yield of the tool trials in 125 mm row width. The dark yellow bars represent the yield of 2017 while the light yellow bars show the amount of harvest in 2018. **Untreated:** No treatment, **Weed free:** Manual weeding,**GFS:** Goosefoot sweep, **NTS**: No-till sweep, **DSK**: Down-cut side knife. Bars with the same letters. Means with the same letter are not significantly different according to Duncan's multiple range test at $\alpha \leq 0.05$. Capital letters represent significance of the trial year 2017 and small letters show the values of 2018. The standard error of the mean (SEM) is represented by the small black bars.

4.3.6 Discussion of Experiment 3

The weed species assessments were dominated by broad-leaf species. This is not the typical case as in Europe cereal cropping is often accompanied by a problematic weed infestation of grass weeds (Melander *et al.*, 2003). A reason for this might a wide crop rotation. Additionally, the trial sites became tillage by ploughing and in 2017 seeding was performed at the end of October.

The shape of a weeding tool influences the weed control result in intra- and inter-row space which is reported also in the study by Pullen and Cowell (1997). Angled or arched blades like the goosefoot sweeps used in this study moved more soil into the intrarow space and at the same time resulted in more soil movement between the rows, thus leading to a better weed control. Compared to that, the flat-shaped no-till sweeps tool is not as effective, especially in the inter-row area and at slow driving speeds.

The lower weed control efficacy at 125 mm row distance might be caused by a too late treatment. Compared to the wider seeding distances, the crop canopy had already started to overlap at the day of treatment. In such situations it was impossible to apply the same aggressive tool adjustment, especially with the goosefoot sweeps blades (to avoid crop damage). Also, proper steering of the hoe was difficult as the clear identification of the crop rows tough. To achieve similar conditions over all trial sites, weeding was done within the same week. Earlier treatment in 125 mm row spaced trials may have solved the mentioned issues.

Weed density 14 days after treatments was significantly lower in the treated plots compared to the untreated controls. This is in line with the findings of Lötjönen and Mikkola (2000). Yet, there is a differentiation in the composition of the newly emerged weeds and the survivors in the hoe treated plots of the different row line widths. Different reasons might have affected these findings. In the 200 mm row width, the wider row space enabled a harsher treatment. As a consequence, the intense sunlight in combination with the dry conditions after the treatments prevented new weed germination. If also taken into account that the ample sunlight induced a growth acceleration of the crop, all the above factors might have suppressed the emergence of new weeds even if the canopy closing was delayed. In the 150 mm row width even if fewer surviving weeds were counted after the treatment, the more favourable micro climatic conditions in the smaller row spacing favoured weed germination, since the new weeds had space, nutrients and light to grow. On the other hand, even at the untreated control plots of the 125 mm seeding distance the quite fast canopy closure appeared to prevent the emergence of new weeds. A major aspect when using hoe tools is crop selectivity. In this study no decrease in selectivity between 200 mm and 150 mm was found. Even though in other studies the recommended row space for hoeing is wider than 150 mm (Lötjönen and Mikkola, 2000; Melander *et al.*, 2003). Concerning selectivity, the no-till sweeps tools on average showed the best selectivity values in all trials. The grain yield results varied per trial. There is a difference of 5.5 t ha^{-1} between the highest yield measured in the 125 mm row distance compared to the lowest one in 200 mm. Such high differences are not common since other studies reported yield differences between the 125 mm and 250 mm row width of around one t ha^{-1} (Rasmussen, 2004; Boström *et al.*,

2012). It should be noted that the 200 mm trials were conducted on an organic farm under the respective conditions. Such fields typically have lower yield potentials compared to the conventional farms. But in 2017 the 125 mm width trial was also on an organic field with a grain yield between eight and nine t ha^{-1}. Similar studies that used smaller variations of row widths (just 200 to 240 mm) have reported small or no yield decrease with the wider rows (Melander *et al.*, 2003). Nevertheless, according to the results of this study hoeing can be used in narrow row spaces and should be further developed. This confirms the findings of Melander *et al.* (2003). The no-till sweeps with the a shape was also favoured by the study of Pullen and Cowell (1997) in wider row width.

4.3.7 Conclusion of Experiment 3

The choice of the right tool for hoeing is highly depending on the soil characteristics and soil condition when the application is feasible and the weed flora composition. Based on the two years of testing in this study, the no-till sweeps performed well over a wide range of soil conditions and row spacing as well as at different operating speeds. Even in heterogeneous soil conditions or in fluctuating driving speeds, due to the soil diversity, the no-till sweeps did not need significant re-adjustments to perform. Especially in the narrow widths of 125 mm and 150 mm, the no-till sweeps outperformed all the tested tools. Therefore, it can be proposed as a robust, general purpose solution for mechanical weed control in conventional farms. On the other hand, the goosefoot sweeps are less universal if conditions are versatile and perform weed control with a lack of selectivity.

4.4 Experiment 4: Adjustment of the camera row guidance systems

The adjustments of the row guidance systems showed large differences in reliability and handling. Even though both systems were performing acceptable under artificial conditions the Naïo-Camera system would have needed substantial improvement to succeed in practice. The Tillett and Hague system, which was originally build for hoe guidance was the better choice.

4.4.1 Naïo

The Naïo camera system initially worked inconsistently and was not able to capture the green lines along the cardboard test. The algorithm was not able to position the pre- set grid of lines on a specific set of green rows. Whenever the cardboard was moved and a different row was exposed to the camera's field of view, the row grid jumped and aligned with the newly visible green line. This confirmed the missing ability of the algorithm to stay locked on a specific set of green lines. Subsequently the steering system followed the new alignment and cause substantial crop damage by overriding one or even two crop rows. Changing the offset to reduce the reaction time of steering enhanced the stability only lightly. Even if the row grid was set to more rows than captured in the camera's field of view, it did not provide the precision needed for hoeing in narrow row distances. This was confirmed in the field tests. In areas with regular crop rows and no weeds in the inter-row space the system performed well for a certain amount of time. Whenever changes appeared the system provided faulty steering commands, lost the correct position and shifted completely to the left or the right.

FIGURE 4.16: Testing the Naïo camera system. **Top left:** Steering signal tests under artificial conditions. **Top right:** 7-inch display as a user interface. **Bottom:** Field testing (B.Kollenda)

4.4.2 Tillett and Hague

The Tillett and Hague system worked from the start in a more reliable way. It was no challenge for the system to steer precisely at various camera positions and different sets of artificial rows. Also, different light sources and their positions did not impact accurate steering. The system provided multiple options to adjust the camera view, the target object colour and the steering unit. The system worked stable under artificial as well as under practical field conditions.

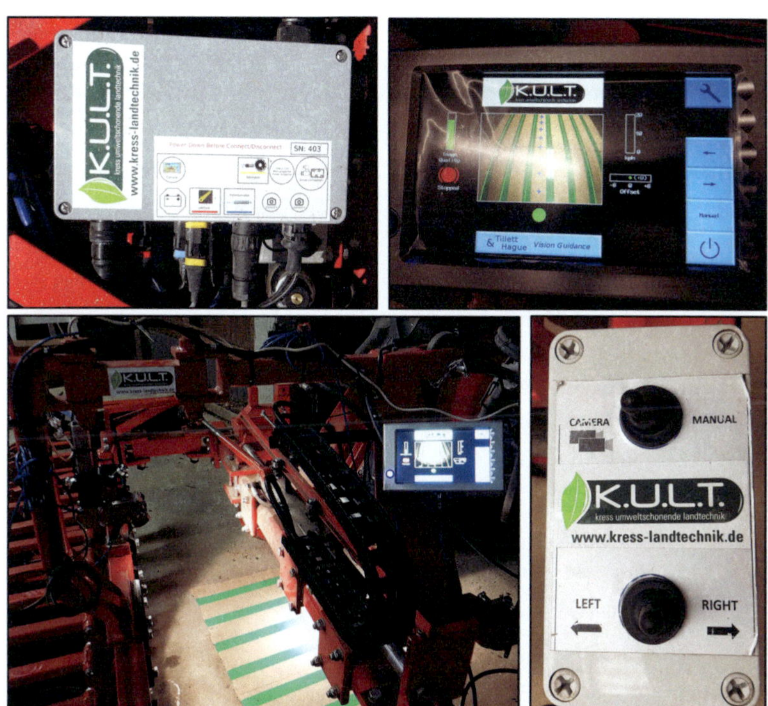

FIGURE 4.17: Testing the Tillett and Hague camera system. **Top left:** Controller unit. **Top right:** Embedded system with user interface. **Bottom left:** Tests with artificial light sources. **Bottom right:** Remote control. (B.Kollenda)

The efforts to adjust a camera row guidance system that was originally designed for another purpose was underestimated. Even though the Naïo system was used already for row recognition there was an important drawback. In contrast to a hoe, the camera system was originally mounted in front of a robot at a position that did not move the same way as the guided hoe tool bar. As a consequence, the software calculating the offset and producing the proper steering signal could not be used without profound changes. Furthermore, the field of view of the camera mounted on the robot was covered to both sides by the robot shell and did not show any extra rows left and right of the current seedbed. The other tested camera row guidance system from Tillett and Hague was originally developed for tractor pulled hoeing equipment. The camera was supposed to be mounted on the tool bar to follow the movements of the tools. The software was already set up for narrow row distances. Nevertheless, it was never tested before as a guidance system in rows smaller than 180 mm. For that reason, the system was useful as camera steered hoeing with less development effort.

4.4.3   Accuracy test

The 3 m hoe is lead by the camera guidance mostly within a distance of 40 mm around the center of the inter-row space. The standard deviation was less than 20 mm. At the beginning the hoe followed the same pattern of lateral movements in every run. Later, the deviation of the first run was changing compared to the second and third one (Figure 4.18).

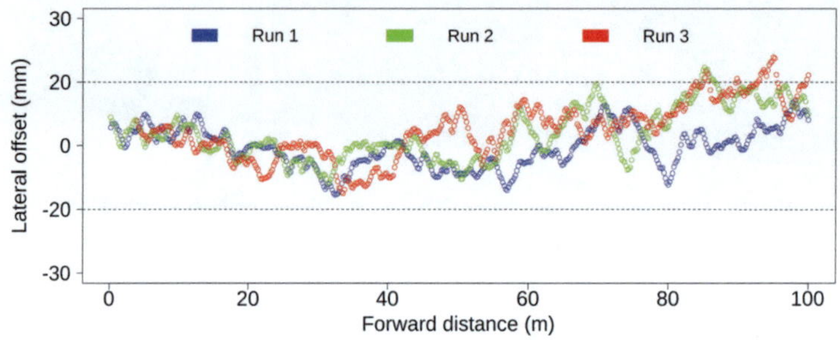

FIGURE 4.18: Lateral offset of the 3 m camera guided hoe. Blue points are the measured lateral position of the first run. Green and red points mark the second and third run.

The 6 m hoe showed a smaller offset between the different runs (Figure 4.19). From the start up to 40 m the first one showed a slightly different pattern compared to the two next runs. Along the last 60 m the hoe guidance almost followed the same track in each repetition. The lateral offset was also within a 40 mm track around the center.

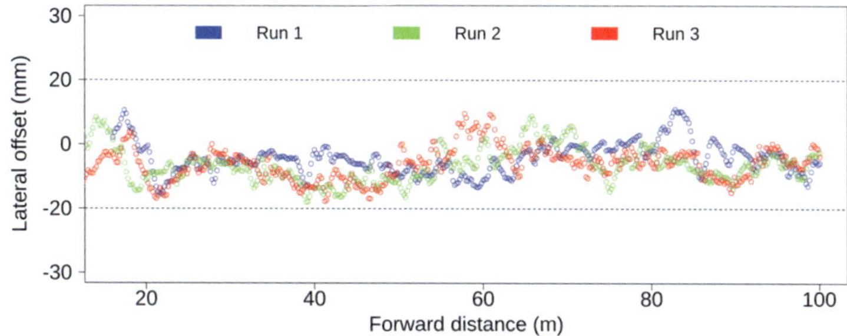

FIGURE 4.19: Lateral offset of the 6 m segmented hoe. Blue points are the measured lateral position of the first run. Green and red points mark the second and third run.

TABLE 4.3: Accuracy of the row guidance recorded by the tacheometer. **SD**: Standard deviation of the lateral distance to the center of the inter-row space

	3m			6m		
	Run 1	**Run 2**	**Run 3**	**Run 1**	**Run 2**	**Run 3**
SD (mm)	18.77	19.16	22.53	18.42	18.05	18.28

4.4.4 Conclusion of Experiment 4

The stability of a row guidance system depends on proper interaction of the sensor system and the steering system. In this study, the hydraulic needed to be adjusted to fit the requirements of the camera row recognition system. The Tillett and Hague system demanded for slow movements of the toolbar in order to reach a stable guidance without overriding rows. As the test runs with this combination were much more stable than the Naïo based row guidance the first hoe prototypes for narrow seed rows were equipped with the Tillett and Hague Vision Guidance. As the accuracy test was conducted in maize, the row shape was different compared to cereal rows. The standard deviation of the row guidance defines the minimum adjustable distance of the tools to the crop rows. Therefore, the crop type to conduct such an experiment is of secondary importance as long as the row guidance can be adjusted to the crop. But cereals show a row pattern where the outer sides are more even, in maize the rather large leaf surfaces which are are orientated towards inside the inter-row space, results in an uneven image for the camera. Further on, the single maize plants are sometimes not evenly centred in the row, which can cause small left and right movements of the hoe steering. The result confirm that these extra movements occur in every run at the same forward distance. Therefore, it is not a guidance error needs to be considered as the normal deviation to keep the hoe alignment with the crop rows. Due to the setup of the hydraulic system, every move in one direction caused a consecutive move in the other direction. Reduction the hydraulic flow might reduce this extra amplitude. The different offset of the hoes can be explained with the camera position. While the 6 m hoe had the cameras in front of the hoe frame with almost no limitation in angle and height, the camera position of the 3 m hoe was not ideal for further developed wide row crop. Results of a study by Tillett *et al.* (2002) in sugar beet were slightly better with a maximum deviation of only 16 mm, while results from (Slaughter *et al.*, 1999) in tomato showed a maximum deviation of 20 mm. Comparability of the results due to the different test method are limited. While (Tillett *et al.*, 2002) checked the lateral offset only every meter in this study the position was recorded more than six times per meter. The higher measurement frequency also leads to a more pronounced scattering of data points, as almost every movement of the hoe was recorded. Therefore, it can be presumed that the hoe system in this study was guided more accurate. Nevertheless, the experiment should be repeated in a trial in cereals to determine the accuracy in narrow spaced rows.

4.5 Experiment 5: Camera guided prototypes

4.5.1 3m hoe prototype

The first tests with the 3 m camera guided hoe prototype were a promising step combining the findings from the tool tests with an automated row guidance. The row guidance precision inside the narrow row distance was at the same level as available equipment on the market for wider rows. Thus, the hoe tools were steered always in the center of two crop rows. The faster driving speed enabled a field efficiency of 1.45 ha h^{-1}. This calculation based on a 100 x 100 m field at 80 % maximum speed and 0.3 minutes per turn on the headland.

4.5.2 Weed species in experiments with the camera guided hoe

The weed assessments for the trials in 2019 showed that only dicotyledonous weed species were present at the trial site HOH. The most abundant species was *L. purpureum* followed by *V. persica* and *V. arvensis.* At IHO the weed composition was also mainly broad leaf weeds and some *A. myosuroides.* The results of the weed community assessment showed that *M. chamomilla* was the most common species followed by *S. media* and *L. purpureum. A. myosuroides* was recorded with an average of 2.8 plants m^{-1}.

TABLE 4.4: Most abundant weeds in the test trial series of the 3 m
camera guided hoe in 2019

Site	Weed Species	Density (m^{-2})
	L. purpureum L.	21.6
HOH	*V. persica* L.	6.1
	V. arvensis L.	1.0
	M. chamomilla L.	24.7
IHO	*S. media* L.	18.0
	L. purpureum L.	15.5

4.5.3 Weed control efficacy

At HOH the inter-row weed control efficacy was similar to the weed free control plot for every mechanical treatment except the fast driven hoe treatment (Figure 4.20. While 3 km h^{-1} and 6 km h^{-1} hoeing speed showed results at the same significance level than the herbicide control plots, the fast version of hoeing was slightly less effective and only at the level of the other hoe treatments. The same pattern of significance levels was determined for the records of the intrarow space. Weed control efficacy was around 90 % on average in the herbicide control plots, an average of 70 % at 3 and 6 km h^{-1} respectively and 53 % average control in the plots hoed with 8 km h^{-1}.

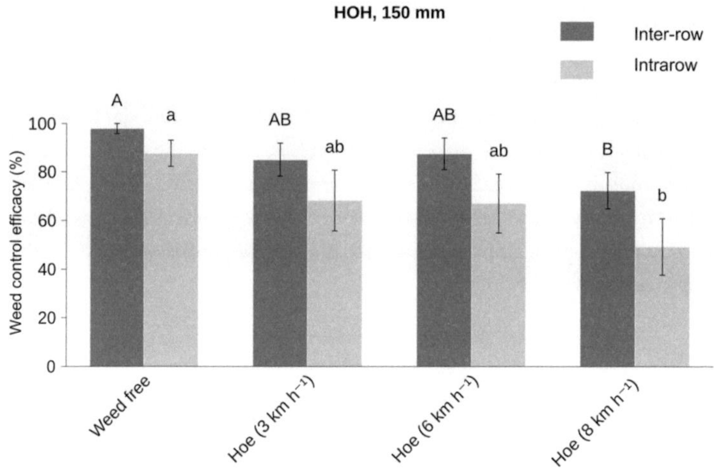

FIGURE 4.20: Weed control efficacy of the 3 m camera guided hoe at HOH. The dark grey bars represents the efficacy between the crop rows while the light grey bars show the intrarow weed control efficacy. Weed free: Herbicides application and manual weeding. Means with the same letters are not significantly different according to Duncan's multiple range test at $\alpha \leq 0.05$. Capital letters represent significance of the inter-row space and small letters show the values between the crop rows. The standard error of the mean (SEM) is represented by the small black bars.

The hoe treated plots at IHO showed an increasing weed control efficacy between 3 km h^{-1} and 8 km h^{-1} for the inter- and intrarow space (Figure 4.21). In the inter-row space, the efficacy was 85 % and 92 % and in the intrarow area 22 %

and 25 % respectively. The test with 6 km h^{-1} did not further increase the weed control efficacy and was on the same significance level as the 3 km h^{-1}. Only 8 km h^{-1} was at the same significance level as the herbicide control plot.

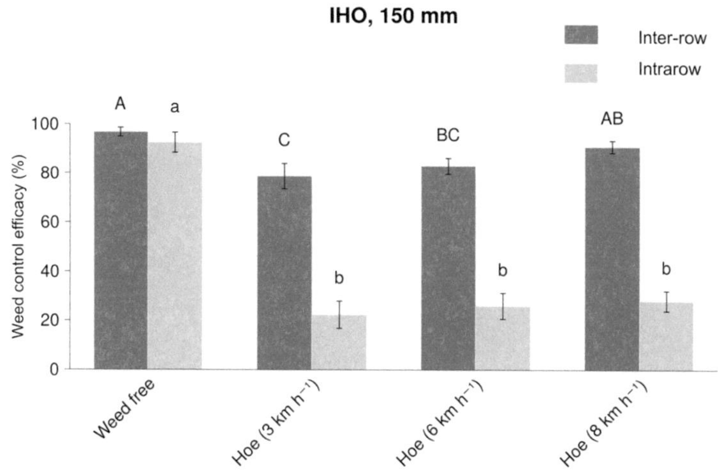

FIGURE 4.21: Weed control efficacy of the 3 m camera guided hoe at IHO. The dark grey bars represent the efficacy between the crop rows while the light grey bars show the intrarow weed control efficacy. Weed free: Herbicides application and manual weeding. Means with the same letters are not significantly different according to Duncan's multiple range test at $\alpha \leq 0.05$. Capital letters represent significance of the inter-row space and small letters show the values between the crop rows. The standard error of the mean (SEM) is represented by the small black bars.

4.5.4 Crop soil cover

The crop soil cover measurements did not show significantly different results, but the values vary up to 16 % around the average. At HOH the treatments lead to an average soil cover of the crop between 28 % and 32 % for hoeing with 3 km h^{-1} and 6 km h^{-1}. Plots treated with 8 km h^{-1} showed an average CSC of 40 %. At IHO the crop rows were less covered with soil compared to the trial located at HOH. The plots treated with medium speed showed the smallest CSC values with an average of 18 %, whereas the crop rows hoed with 8 km h^{-1} were covered by 25 % of soil (Figure 4.22).

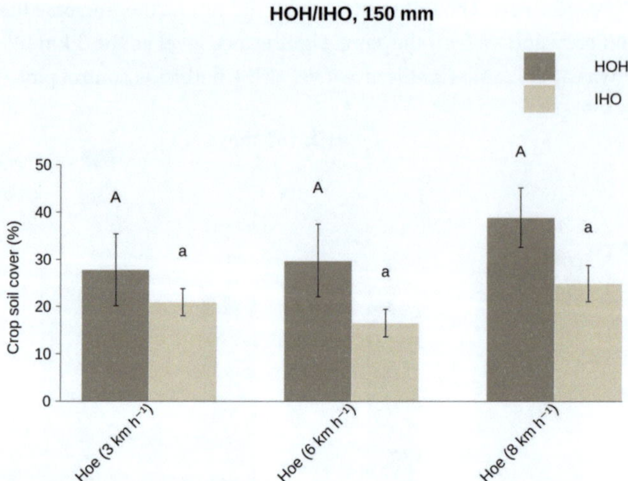

FIGURE 4.22: Crop soil cover of the test trials with the 3 m camera guided hoe. Bars with the same letters are not significantly different according to Duncan's multiple range test at $\alpha \leq 0.05$. The standard error of the mean (SEM) is represented by the small black. bars.

4.5.5 Grain yield

The yield in the 3 m camera guided hoe tests did not differ significantly at both trial sites across all treatments (Figure 4.23). The average yield was around 7 t ha^{-1}. At HOH the untreated control showed a tendency to lower grain yield and the highest yields were recorded in the permanent weed free plots with 7.6 t ha^{-1}. The hoe treatments did not impact yield negatively even at a speed of 8 km h^{-1}.

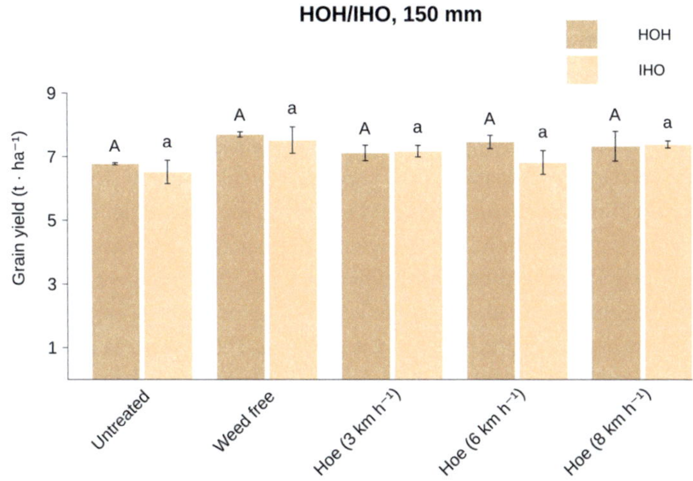

FIGURE 4.23: Grain yield of the test trials with the 3 m camera guided hoe. Bars with the same letters are not significantly different according to Duncan's multiple range test at $\alpha \leq 0.05$. The standard error of the mean (SEM) is represented by the small black bars.

At IHO the average yield was also around 7 t ha^{-1} but results vary slightly more. The herbicide control plot showed the highest yield values. At this site the 6 km h^{-1} operation speed lead to slightly smaller amount of grain yield while the 8 km h^{-1} was at the same level as the herbicide control.

4.5.6 Discussion of the results achieved with the 3 m camera guided hoe

The reduced weed control efficacy at HOH in the 8 km h^{-1} hoeing treatment can be explained by a dry and compacted soil conditions that prevented the tools to penetrate into the topsoil. It was observed in video records that the hoe parallelograms started to jump during fast operating speed and did not perform along the whole plot. This was even worse at IHO which lead to a poor result of weed burying in the intrarow space. Even during the fast-driven treatment, the records did not provide significant better weed reduction. This result was mainly caused by the wet spring which reduced the time-window of hoeing to a few days only. Even at the specific days where treatments were performed the soil conditions were not ideal and often slightly too wet at the time of hoeing. The soil condition also explains the crop soil cover values. Whereas at HOH big soil conglomerates were moved in the intrarow space and lead to a higher reduction of leaf cover, the crop soil cover at IHO was too small to reduce the weed infestation until canopy closure. The grain yield showed that the remaining weeds in the intrarow space did not impact the yield potential. The operating performance of the 3 m camera steered prototype was satisfying. The compact design of the construction lead to a relatively small distance between the rear axle of the tractor and the tool bar which reduces the movement of the tools in lateral directions in case the tractor is miss-aligned with the crop rows. The 200 mm maximum compensation steering limits the ability of the hoe to properly follow seedbeds with curvy shapes. The camera position between the tower arm and the fixed frame reduced the options of adjusting the camera field of view. At smaller crop growth stages this was not problematic but if treatments were delayed it was not the best position and just a compromise between the proper field of view and the height. An enlargement of the pivot arms might be a solution to have a wider space between the fixed and the flexible part for the camera adjustments. This would also enable a longer compensation distance for the hydraulic steering.

The field efficiency of the camera guided hoe was almost doubled compared to a manually steered hoe due to increased operating speed. Similar results are reported by Kunz *et al.* (2015) in sugar beet and soybean. Nevertheless, for larger scaled agriculture a 3 m hoe is too small as the operating time per ha does not fit the requirements to conduct weed management in a short time window. As mentioned in the introduction an enlarged toolbar can only be possible when also enlarging the sowing equipment. A solution for that was the 6 m segmented hoe of which tests are described in the following section.

4.5.7　6 m segmented hoe

Compared to the 3 m prototype, the 6 m hoe has doubled the field efficiency and did not lead to a major delay by turnings at the headland. Also, the preparation time from the folded status for transportation to open status for field use did not take much longer than the preparation of the 3 m hoe. The whole process took around 5 minutes. The hoe was more susceptible to reach the steering limit of the row guidance compared to the 3 m version in case of miss-alignment with the crop rows due to the enlargement. Nevertheless, an acoustic signal informs the operator in if damage of the crop is at risk.

4.5.8 Weed control efficacy and crop soil cover of the 6m hoe

The weed control efficacy at 4 km h^{-1} speed resulted in an average of 90 % weed control in the inter-row space relative to the untreated control (Figure 4.24).

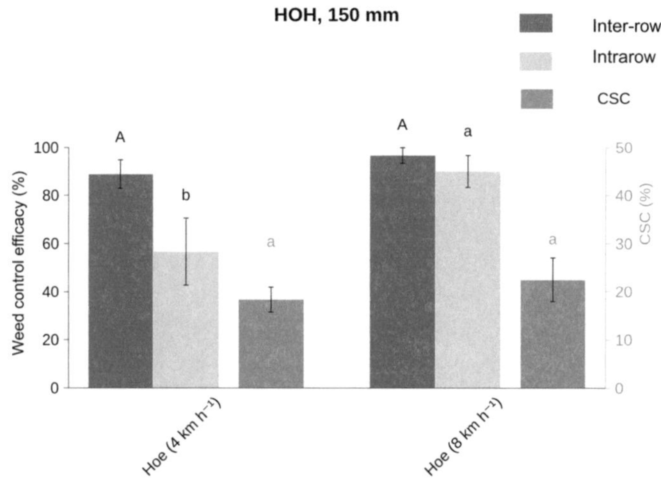

FIGURE 4.24: Weed control efficacy and crop soil cover recorded in the test trial for the 6 m segmented hoe. The left y-axis shows the weed control efficacy and the right y-axis shows the amount of crop soil cover. The black bars represent the efficacy between the crop rows while the light grey bars show the intrarow weed control efficacy. The dark grey bars show the amount of crop soil cover. Means with the same letters are not significantly different according to Duncan's multiple range test at $\alpha \leq 0.05$. Capital letters represent significance of the inter-row space and small letters show the values between the cop rows. The small grey letters are for the crop soil coverage values. The standard error of the mean (SEM) is represented by the small black bars.

For the intrarow area the weed control efficacy was in a range between 43 % up to 75 %. The test with 8 km h^{-1} showed an inter-row weed control similar to 4 km h^{-1}. Significantly better control in the intrarow space was observed. The CSC was at the same significance levels for both velocities but usually higher in the plots with faster treatment speed. The crop cover by soil was a little bit higher at 8 km h^{-1} with an average of 22.5 % compared to 20 % in plots treated with 4 km h^{-1}. Nevertheless, the statistical analyses showed no significant differences in this case.

4.5.9 Grain yield achieved by the treatments with the 6 m hoe

The yield results showed that the untreated control plots yielded in average around 4 t ha^{-1} while in the hoe treated plots operated at 4 km h^{-1} in average 6 t ha^{-1} were harvested. The yield in the 8 km h^{-1} hoe treatments had one ton more per ha with a mean of 7.1 t ha^{-1}. The yield in the three treatments were significantly different from each other (Figure 4.25).

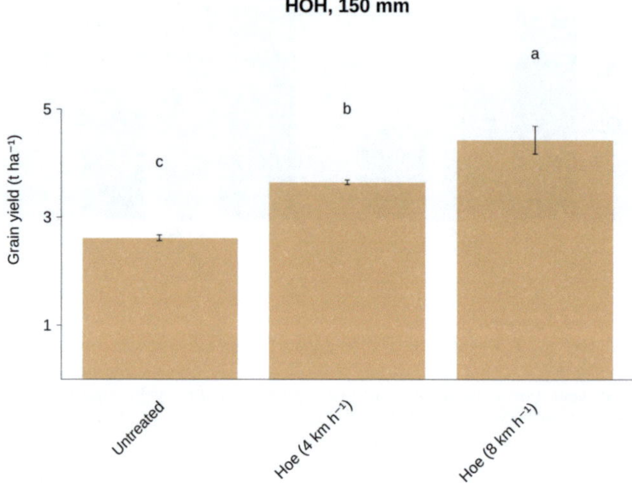

FIGURE 4.25: Yield of the summer barley trial treated with 6 m segmented hoe. Means with different letters are significantly different according to Duncan's multiple range test at $\alpha \leq 0.05$. The standard error of the mean (SEM) is represented by the small black bars.

4.5.10 Discussion of the results achieved with the 6 m hoe

The weed control efficacy results of the field test with the 6 m segmented hoe were better in the intrarow space compared to the other hoe trials. A factor that lead to the high weed control might be due to the relatively low and uneven weed infestation in the trial field. The most abundant species was *C. arvense*. The slow driving speed did not move enough soil to uproot and cover the thistle plants. Furthermore, the trial was conducted in a field where chemical plant protection was applied in the seasons before. Even if the crop soil cover was a bit higher in the faster treatments the crop seemed to recover from that burying and benefited from the reduced competition of weeds. A significantly higher yield was achieved in 8 km h^{-1} treated plots. During operating, it was important to keep the correct position between the two beds when no RTK-GNSS-steering is available. The wider distance of 2.4 m from the rear axle of the tractor to the tools lead to wider movements of the hoe even when the tractor had only small deviation from the center row. At this point, the 3 m hoe has a better range of accuracy. For up- and downhill runs in sloped fields the camera guidance-system provide a manual offset adjustment at the user interface to prevent side slopes that may cause crop damage. The camera position in front of the hoe was advantageous: more rows could be observed due to a higher position of the cameras. Further on, the seedbed left and right of the tractor was large enough to mount the cameras towards the outer sides to reduce shade influences in the field of view caused by the tractor cabin. The development of hoes with wider working widths is limited by undesired flexibility and twisting of the hoe frame relative to the tractor. To solve this issue, a circle compensation movement around the center of the hoe needs to be developed in future. Hoes with segmented row guidance systems were already introduced previously. But those models differ in an important detail: Other concepts consist of three sections where the middle section is mounted at the front hitch of the tractor (e.g.Tillett (2005)). Hence, the critical distance between the segments at the back of the tractor is eliminated as they are one seeding strip apart from each other. As these other segmented hoes are used in wider row spaces with more lateral flexibility it needs to be further investigated if such a setup is also feasible for narrow seeded cereals. The 6 m prototype developed in this study also needs some improvement have a model ready for the market. Even though the anti-collision system worked without any malfunction it needs to be supplemented by a redundant backup. Also, a touch sensor should be installed in case the ultrasonic sensor stops working. The foldable frame can be replaced by a lighter and a more compact design to reduce the weight during transportation and shorten the space between turning point of the tractor and the hoe tools.

4.6 Integrated weed management (IWM) trials

4.6.1 Weed control efficacy

The combination of mechanical and chemical weed control methods seemed
to complement each other positively and lead to a higher total weed control
efficacy (Figure 4.26). In 2018 at IHO 1 in the herbicide control plots around 95 %
of the weeds were removed. At IHO 2 the herbicide control showed some lack of
efficacy and reduced the weed infestation in average to just 80 % in the intrarow
space.

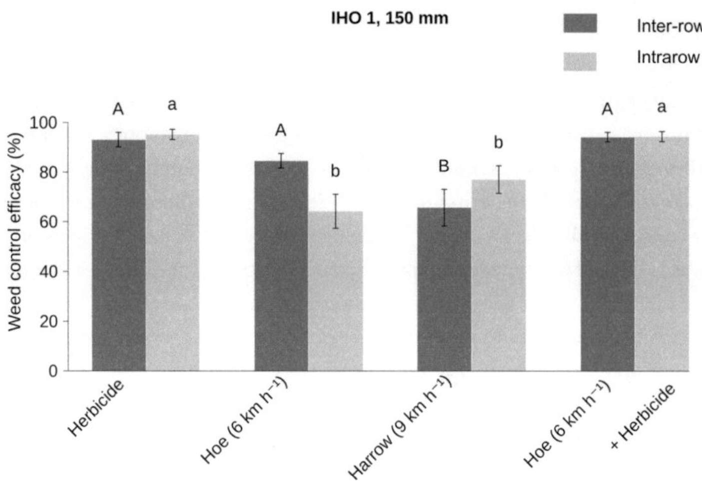

FIGURE 4.26: Weed control efficacy for inter- and intrarow weeds at
IHO 1. The dark grey bars represents the efficacy between the crop
rows while the light grey ones show the intrarow weed control efficacy. Herbicides: Means with the same letters are not significantly
different according to Duncan's multiple range test at $\alpha \leq 0.05$. Capital letters represent significance of the inter-row space and small
letters show the values between the cop rows. The standard error
of the mean (SEM) is represented by the small black bars.

Between the crop rows only the harrow showed significantly lower weed
control. In the intrarow space of the hoe and the harrow treatments the weed
control efficacy was at the same level at both sites (between 50 % and 80 %).
But there was a tendency of better control efficacy in the harrow plots. The

combination of hoeing and herbicide application was at the same level of weed control with the herbicide only plots (Figure 4.27).

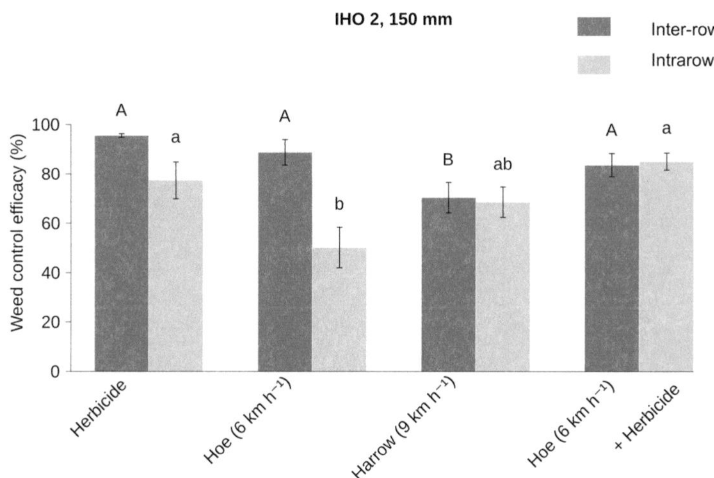

FIGURE 4.27: Weed control efficacy for inter- and intrarow weeds at IHO 2. The dark grey bars represents the efficacy between the crop rows while the light grey ones show the intrarow weed control efficacy. Herbicides: Means with the same letters are not significantly different according to Duncan's multiple range test at $\alpha \leq 0.05$. Capital letters represent significance of the inter-row space and small letters show the values between the cop rows. The standard error of the mean (SEM) is represented by the small black bars.

In 2019 the herbicide treated control plots showed weed control efficacy levels of around 98 % in the inter-row space and at both sites around between 82 and 98 % between the crop rows (Figure 4.28). The use of the harrow alone lead to reduced weeding success at maximum levels of 65 %. While at HOH the weed control in the intrarow space was just around 40 %, at IHO it was higher than between the crop rows (up to 70 %).

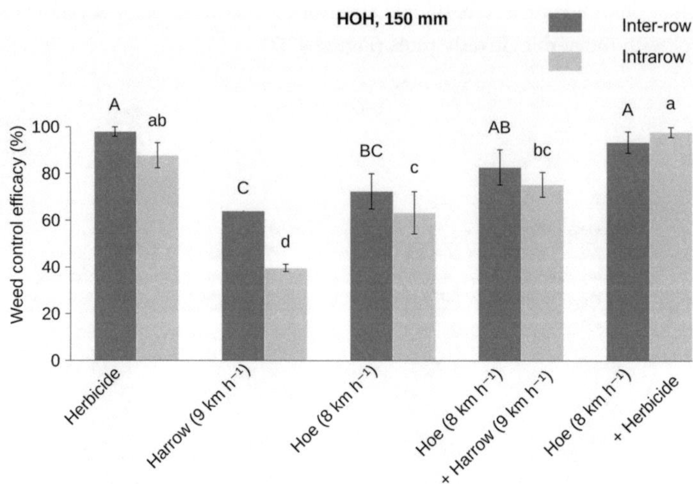

FIGURE 4.28: Weed control efficacy for inter- and intrarow weeds at HOH. The dark grey bars represents the efficacy between the crop rows while the light grey ones show the intrarow weed control efficacy. Herbicides: Means with the same letters are not significantly different according to Duncan's multiple range test at α ≤ 0.05. Capital letters representing significance of the inter-row space and small letters shown the level for values between the cop rows. The standard error of the mean (SEM) is represented by the small black bars.

In the intrarow space one time hoeing lead to a weed control efficacy at the same level as the harrow (HOH) or even significantly better at IHO. The weed control in the intrarow space of the hoe alone was significantly lower at both sites compare to the harrow. All combined treatments at both sites showed the same significance level as the herbicide control. But only the combination of hoeing and herbicide application led to significantly better results in the inter-row space than the hoe alone. Here the total weed control efficacy was around 95 %. Only in the intrarow space at IHO hoeing and harrowing showed an enhanced weed control efficacy (Figure4.29).

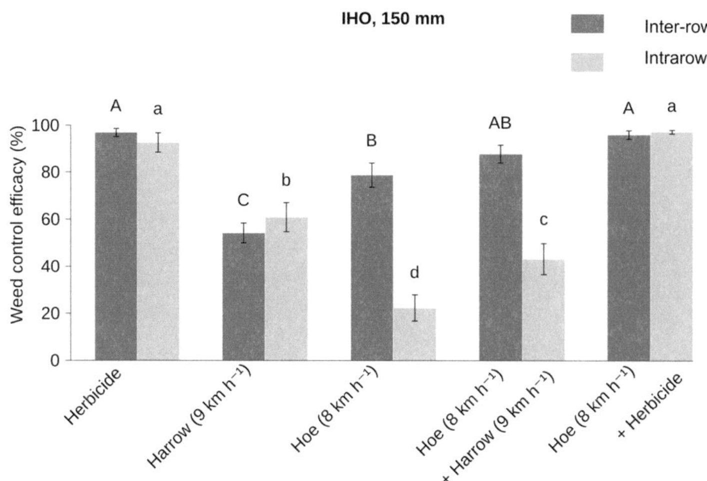

FIGURE 4.29: Weed control efficacy for inter- and intrarow weeds at IHO. The dark grey bars represents the efficacy between the crop rows while the light grey ones show the intrarow weed control efficacy. Herbicides: Means with the same letters are not significantly different according to Duncan's multiple range test at $\alpha \leq 0.05$. Capital letters representing significance of the inter-row space and small letters shown the level for values between the cop rows. The standard error of the mean (SEM) is represented by the small black bars.

4.6.2 Grain yield

Grain yield harvested in 2019 was less than 7 t ha^{-1} in the untreated control plots and 7.5 to 8 t ha^{-1} on average in the herbicide treated plot. At HOH the yield levels did not differ significantly from each other. But at IHO the hoe treatment in combination with a herbicide application provided significantly better yields than the untreated control, the harrow treatment and the hoe alone. Also, some plots in HOH showed the highest yield for the combination of hoe and herbicide treatment. Yield in harrow alone treated plots was at both trial sites smaller. The hoe in combination with the harrow brought a slightly higher yield (Figure 4.30).

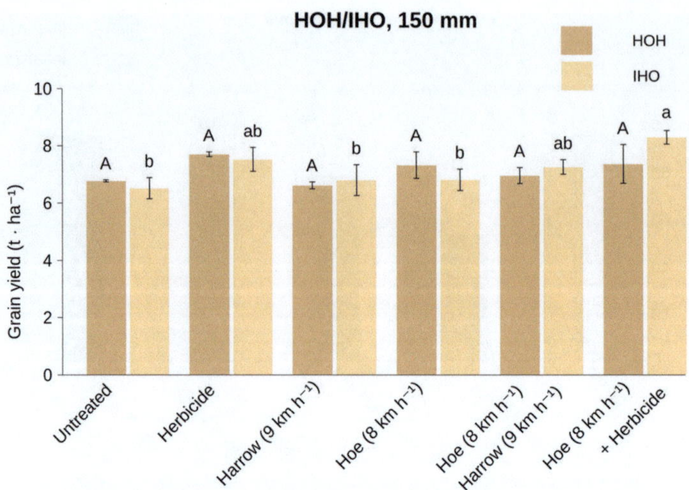

FIGURE 4.30: Grain yield of the weed management trial in 2019. Means with different letters are significantly different according to Duncan's multiple range test at $\alpha \leq 0.05$. The standard error of the mean (SEM) is represented by the small black bars.

4.6.3 Discussion of Experiment 6

At both sides the most abundant weeds were all dicotyledons which are not problematic to control by hoeing. As *L. purpureum* was not sufficiently removed in the herbicide control plots by the application the weed control levels were lower than 95 % at HOH. The remaining weeds needed a manual pulling after the weed control efficacy record. The results pointed out that hoe treatments alone did not provide a full reliable weed control in all cases. Additionally, the highest competition is caused by weeds grown in the intrarow space (Melander *et al.*, 2012) where the weed control performance was often poor. Therefore, the combination of different weeding approach can be hedging. The results of the weed management trials confirmed that hoeing and harrowing in combination enhance the weeding performance. This agrees with the results of a study by Melander *et al.* (2005). The smaller weed control performance of the harrow in the inter-row space is also reported by other studies (E.g. Pullen and Cowell (1997)).

The test trial of the tools in Experiment 3 showed that weeds that emerge after weeding and before canopy closure did not reach a competitive growth stage to influence the yield. Except some species that can climb up the crop and grow above the canopy (e.g. *P. convolvulus* and *G. aparine*). These weeds can produce seeds and also affect harvest operations. This agrees with findings by (Wilson and Wright, 1990). As they can emerge in the inter-row space, a herbicide application of the whole plot was conducted. In all trials where hoeing and herbicide application was combined, the weeding level of the inter- and intrarow space was equal and also not different from the herbicide control plots.

If weeding is possible at early weed growth stage harrowing can be a very effective measure controlling them as driving speeds of 9 km h^{-1} are possible. Also, the efficacy on intrarow weed infestation can be higher compared to a hoe treatment. Nevertheless, hoe tools can also control further developed weeds in case suitable conditions only occur later in spring. The decision which combination can be chosen for successful weed management first depends on the operating method. In case non-chemical treatments are the only choice due to organic certification the combination of harrow and hoe is beneficial especially in case of fine-crumbly soil conditions. To enhance the result of the mechanical weed treatments the weed infestation after crop emergence should be as small as possible. This might be achievable by cultural and preventive measures. For example, wider crop rotations and a reduced part of winter cereals in combination with a late sowing date can be helpful to prevent a major grass-weed infestation. Use rotating tillage and fault seedbed preparation enables effective options to reduce the emergence of perennial weed species before crop emergence. To also suppress weed emergence between cropping seasons the use of cover crops should be considered. Mulching of the cover crop and the remaining mulch can enlarge the effect after crop emergence.

In case herbicide applications can be part of the weed management strategy a successfully weed control might be possible with a reduced amount of passes by a band-spray application where hoeing and spraying equipment is combined. This can be a way to reduce herbicide input in conventional cropping systems with the same efficacy than a full-dose application.

4.7 Summery of the results

Summarising the most important results of all experiments to have a better overview should be done in this section. The weed control efficacy and grain yield results of the no-till sweeps are shown, in 150 mm row width and under highest velocity conducted in the specific trial site. The yield of the trials conducted in winter wheat showed similar grain yield. The weed control efficacy varied from trial to trial. In the experiments were the camera guidance was used the weed control was higher. The best control was achieved in combination with a herbicide application. Details are presented in Table 4.5

TABLE 4.5: Summery of the weed control efficacy and the grain yield of the fast driven no-till sweeps in all Experiments. [1] 3 m camera guided hoe, [2] 6 m camera guided hoe (*Trial was conducted in spring barley), [3] 3 m camera guided hoe and herbicide application at HOH site (as the 8 km h^{-1} was not successful due to compacted soil conditions the results of the 6 km h^{-1} are presented), [4] 3 m camera guided hoe and herbicide application at IHO site

Experiment ID	Velocity km h^{-1}	WCE inter-row %	WCE intrarow %	Grain yield t ha^{-1}
2	3-4	70	68	-
3	6-8	88	70	7-9
5a[1]	6-8	85	65	6.5-7.5
5b[2]	8	94	89	4.5*
6a[3]	8	90	95	7.2
6b[4]	8	92	96	8.3

Chapter 5

General Discussion

Inter-row hoeing is a measure that highly depends on the correct preparation of the crop. Hoeing can only conduct if the seeding is done adequate, concerning the spacing of the crop rows to each other. Therefore, seeding equipment with flexible coulters or ones with worn bearings can not be used in combination with this measure. According to this study successful hoeing in narrow seeded cereals depends on some circumstances that need to be fit during the treatment:

- the soil condition and the timing

- the tool shape and the alignment to the rows

- proper guidance along the crop rows

To perform successful weed control with a hoe, the soil needs to be relatively dry and loose. This allows efficient transport of soil from the inter-row space into the intrarow space to bury weeds and thus negatively impact their development (Baerveldt and Ascard, 1999). But a too dry topsoil can be also problematic as in such situations hoeing generates dust that, once deposited on crop leafs, impacts growth negatively (Pullen and Cowell, 1997). The choice of row-width also is a decisive factor: The narrower the rows are sown, the smaller is the time window for mechanical weed control treatments. On the other hand, the more narrow row widths showed a decrease in new emerging of weeds after hoeing, most probably due to a faster canopy closure and the higher crop density. This agrees with the findings of Fahad *et al.* (2015) in their study comparing row distance of 230 mm to 110 and 150 mm in wheat. Further on, timing of the treatment is very important as late measures can be ineffective against well-developed weeds and, on the other hand, small weeds treated early may survive in the topsoil (Pullen and Cowell, 1997). Compare to Pullen and Cowell (1997) where the hoe was equipped with protection shields, the aim in this study was to allow sufficient transport of soil from the inter-row space in the crop rows to cover small weeds.

It is commonly known that with most blades weed control efficacy is a function of driving speed (Melander *et al.*, 2003; Kunz *et al.*, 2016). This is confirmed by the results from this study which indicate that there is a trend that treatments with a higher speed increased mechanical weed control efficacy. A hoe with an automatic row guidance system can conduct hoeing much faster and more precise especially in narrow row distance (Pullen and Cowell, 2000, 2006; Kunz *et al.*, 2015). According to Prince *et al.* (2012) a hoe equipped with automatic row guidance can be an important tool to control herbicide-resistant weed biotypes. However, such a set up is expensive and not yet affordable for smaller farms until the costs for camera row guidance systems decline.

As hoeing can control already further developed weeds and also tap-rooted species, it is often argued that perennial weed species can be further spread by translocation of their rhizomes. In this study, hoeing was performed flat in only 20 mm depth. This should not impact the spread of perennial weeds. Therefore, the weeds were just cut off slightly below the soil surface. On the other hand, in some of the trials well-developed *C. arvense* survived the hoe treatment even when a plant emerged in the middle of the inter-row space. This underlines the importance of regular sharpening of the tools to ensure proper cutting of weeds with more robust tissue. If perennial weeds occur in the intrarow space they will most probably survive because burying them with soil will not stop them from further growing (Lötjönen and Mikkola, 2000). Also, the biomass ratio of root (high) to shoot (low) allows perennial weeds to regrow very fast, as they can compensate for lost shoots by mobilising starch, water and nutrients from the rhizomes. Therefore, they can only sustainable reduced by cultivation measures such as deep tillage using a plough (Melander *et al.*, 2012).

As mentioned before a key factor for hoeing as an alternative to herbicide applications is the field efficiency. Only when hoeing can be conducted with similar operating speed and comparable working width like sprayers, mechanical weeding will become a competitive option. The study addressed this by the aim to develop a hoe with camera guidance for narrow row space. Building the prototypes in 3 m and 6 m working width was successful by the combination of a highly developed camera row guidance system, a very fine adjustable steering frame and a well-tested hoe tool, usable over a wide range of soil conditions. Treatment speed in this study even with the 6 m segmented hoe was limited and might be still slower than a herbicide application with equipment with large treatment widths. The 6 m hoe can be seen as an example that the principal of a segmented hoe can enlarge the working widths successfully. Another drawback is the setup time of a hoe that can be time-consuming due to necessary adjustments to crops and soil conditions. Nevertheless, this is similar to the time needed to fill a sprayer with water, mixing the herbicides and checking the nozzles. A hoe, once adjusted does not need a refill nor runs to refill or to empty the tank afterwards.

The reasons for a faster move from chemical to mechanical weed management are diverse. Missing weed control efficacy of hoeing is not the major problem that prevents wider introduction. Some mechanical treatments in this study showed similar weed control levels as herbicide treatments. But compared to herbicide spraying, the adjustment of the tools as well as the correct interpretation of the weeding results cannot be communicated as easy as on a herbicide label as these aspects vary substantially from site to site. Hoeing adjustment is a task that needs training and experience to be able to value the weed control success during the treatment. Further on, herbicide applications can be conducted under a wider range of weather and soil conditions and herbicides normally have a rain fastness of a few hours. The weed control efficacy of the hoe treatments tested in this study was higher compared to other studies (Pullen and Cowell, 1997; Lötjönen and Mikkola, 2000). Nevertheless, compared to a herbicide applications the weeding performance in the intrarow space was often poor. Therefore, the major challenge of hoeing is the right selection of prophylactic measures in combination with mechanical weeding passes (Melander *et al.*, 2005). A reduced weed infestation achieved by a soil turning tillage measure, followed by a stale seedbed preparation can help to improve the prerequisite for hoeing. Melander *et al.* (2005) recommended to use a harrow as pre-emerged weeding. This measure can further reduce the weed pressure in the cropping season and also eliminated the selectivity problem of the harrow. Hoeing can be used post-emergence multiple times as the main direct measure, in case of low weed infestation in narrow seed rows most probably without any herbicide. Experiment 6 just partly answered the question if hoeing can play a major role in an integrated weed management approach. A major limitation was that the trials conducted in that study were not large enough to represent practical situations. Thus, with that small scale trials, it was easy to conduct hoeing within the time window were the soil condition was suitable. Further on, the fields on which the trials were set up weeding was done by herbicides in the years before and preventive weeding methods were not part of the experimental question. The weed infestation was always within certain limitation and not out of control.

According to a study by Bastiaans *et al.* (2008) the combination of multiple weed control methods will lead to a more diverse and complex cropping system. Thus, agricultural research is in duty to design, test and communicate clear evaluations of alternatives to herbicides.

Chapter 6

Conclusion and Outlook

Mechanical weeding was used for thousands of years to control weeds in crops. The chemical weed control era is, compared to that, just a very short period in the history of agriculture. The fact that herbicides are the major part of all weed control techniques in developed countries might change, as a result of controversial discussions in modern societies. Simultaneously, the high productivity level which is partially linked to chemical plant protection needs to be stabilised and further increased. This study was conducted to address the question on how to complement or even to replace chemical weeding by alternative methods. The main goal was to establish high precision mechanical weed control in cereals by hoeing. This study confirmed that efficient hoeing is feasible with the developed row guidance system even in cereals drilled in narrow seed rows. With adapted no-till sweeps hoeing can be performed across a wide range of soil conditions and different driving speeds. A certain limitation in the conducted trials was the absence of grass weed species. Grass weeds are the major problem in some areas in Europe if they occur in high densities in cereals. Their presence can sometimes limit the success of chemical weed control due to the selection of herbicide resistant weed biotypes. So in future, it is important to test the new hoeing system against grass weeds. Based on the results on broad-leaf weed control it is expected that grass weeds will be controlled at similar levels. The row guidance system allowed higher driving speeds, more exact guidance of the tools along the crop row and was able to enlarge the field efficiency of hoeing. The results of both prototypes showed that sufficient weed control can be achieved, especially in combination with other weeding methods. The adopted hoe tools did not damage the crop and grain yields were not reduced by the hoe treatments, when compared with manual weeded control plots or to herbicide applications. The major challenge is to compensate for the limited weed control reliability by the optimisation of other measures before crop emergence. At the end of Chapter 1, the aims are presented that were used as a goal for this study. The first aim (1) was testing the selection of modified hoe tools for narrow seed rows. This goal was achieved. Goosefoot sweeps, no-till sweeps and the down-cut side knife at row spacing of minimum 125 mm were selected. Where the no-till sweeps were most selective to the crop in combination with weed control levels of more than 90 % per pass. Achieving high levels of weed control was Aim 2. Aim 3 was the development of a camera based row guidance. The successful adaption and the test of the camera row guidance of Tillett and Hague in 150 mm row width was a key milestone to further develop hoeing in cereals. The outcome from the hoeing tool tests and the adaption of the camera system were the prerequisites for designing a 3 m and a 6 m wide automated hoe prototype which was Aim 4. In the final experiments, hoeing in narrow seeded cereals was tested as part of an integrated weed management, in combination with other mechanical and chemical methods. The tests showed that together with a herbicide application, or a harrowing treatment, the weed control efficacy achieved almost full control in the intra- and inter-row space. The conclusions form the experiment are an important step towards the development of hoeing as a large scale weed management solution, also for conventional farming. The possibilities and limitations were discussed and can be seen as the base for further research. As a next milestone hoeing in narrow seeded cereals should be implemented as part of a crop rotation under practical conditions in larger fields.

6.1   Outlook

Further research with the focus on the reliability of hoeing, as part of a cropping system in narrow seeded cereals, should include the following aspects:

1. Numerous test-trial sites with a wide range of crop rotations, soil types and weather conditions need to be conducted to have representative for agronomic practices in Europe.

2. Beside a pure hoeing measure the combination of hoeing and other direct weed control measures should be further tested to achieve the highest possible weed control and investigate the most successfully combination for specific regions, soil types, and crops.

3. To gain best circumstances, in term of the some times reduced weed control efficacy of hoeing, a comparison should be done of different preventive

and cultural methods, to identify a cropping system which can be based on hoeing measures without raising a major weed infestation problem.

4. To enable hoeing also as a economically competitive measure different amount of measures per season should be tested to find the minimum amounts of passes and gain acceptable weed control performances.

Trials in the study were conducted on sites with similar soil types. But the soil condition was different according to the weather the days before treatment. The weed control efficacy of hoeing partially depends on the present soil condition. Even though the no-till sweeps can be seen as a quite universal tool, further tests in sites where all prevailing conditions are different might be necessary to prove the reliability of the proposed measure. Further on, the weed composition can be similar on more diverse sites. This can also affect the weed control success, and in consequence, remaining weeds can limit the yield potentials. Consecutive trials also should consider sites with high levels of grass weed infestation to see if hoeing can be a managing solution on that problem. Combining a hoe treatment with other measures e.g. herbicide spraying or harrowing, might enhance the reliability under different conditions. The combination can be done as separate passes for each measure or as attachments on the hoe. Means, herbicide applications can be conducted at the same time by a hoe equipped with a band-spray unit. Harrow elements, as tested in Experiment 2, can be further developed. If they would have similar tensions as a spring tine harrow they can be mounted as elements between the inter-row tools on the tool bar. Hereby, it is important that these harrow elements working in front of the hoe tools in the intrarow space. In such a setup they can control all small and less developed weeds between the crop plants in a first step and the hoe tools can further on, bury them with soil. The prevention of high weed infestation levels might enhance the weed control success of each hoeing pass. Therefore, crop rotation, tillage and seeding time need to be optimised to achieve small amounts of weed density afterwards in the crop. A test of hoeing at different crop growth stage can help to find the optimum weed control time. The proposals in the conclusion and the outlook section should make clear that the study was able to reach the main goal to successfully hoe in narrow seeded cereals. But more research is necessary to observe long-term effects on the weed population and to establish this measure as a possible standard weed management strategy in cereal production.
The hoe prototypes also need to be improved. The 3 m hoe might profit from longer pivot arms to have a wider field of view for the camera, in order to track

more crop rows. Additionally, this might allow a wider steering way to allow more offset of the tractor to the crop row, without crop-damaging. The synchronous cylinder, which is quite large and heavy, most probably can be replaced by a smaller version to reduce the total weight of the hoe system. To further stabilise tracking under unfavourable sunlight by reducing shade patterns in the field of view of the camera, different types of illumination should be tested under daylight condition. As cereals can be quite tolerant to over-passing by machinery equipped with wide tyres in the stage when mechanical weeding is conducted, a setup where no driving track is established should be tested in order to increase the yield potential. Therefore, the cone-wheels in front of the fixed frame need to be replaced by wider supporting wheels equipped with tires to safe the crop row. With the 6 m hoe, this can also allow multiple passes with an offset to the pass before in order to prevent soil compaction at a specific area in the field.

Based on the 6 m segmented hoe, tests can be done with more segments to enlarge the working width and also the field efficiency. Hereby, the possible increasing flexibility of the complete hoe frame needs to be taken into account. This horizontal offset can lead to a miss-alignment of the outer tools and need to be addressed by a compensation movement. Besides that, supervision of the single tools can be problematic and should be addressed by an alert system. Otherwise, unnoticed clogging of tools will cause crop damage. Another optimisation approach is the site-specific automated adjustment of the hoeing aggressiveness according to the amount of weed infestation or the weed species morphology. This can be either achieved by a single tool lifting mechanism or by an adjustable angle of the tool bar. While the possible adjustment of the tool aggressiveness is easy to implement this approach need to have a camera system able to distinguish not only between crop and weed plants but also recognise which type of weeds are in the field of view. By the use of artificial intelligence e.g. a convolutions neuronal network a highly sophisticated algorithm might be able to set predefined adjustments to meet specific weed infestations and safe the crop at clean patches. Nevertheless, these suggestions might increase the attractiveness of mechanical weeding, all means of further developing need to be proofed on their effect of increasing production costs and if the products are still affordable for farmers.

The change in weed management from herbicides to alternative weeding measures, will impact agricultural production as a whole. A sustainable and safe food production can only be secured when such alternative weeding methods can be fundamentally optimised. Only testing means previously considered as not feasible, will result in the discovery of new technologies and systems that will drive further progress in weed management.

Bibliography

Anderson, R. L. (1994). Characterizing weed community seedling emergence for a semiarid site in Colorado. *Weed technology*, 8(2):245–249.

Åstrand, B. and Baerveldt, A. J. (2005). A vision based row-following system for agricultural field machinery. *Mechatronics*, 15(2):251–269.

Baerveldt, S. and Ascard, J. (1999). Effect of Soil Cover on Weeds. *Biological Agriculture & Horticulture*, 17(2):101–111.

Bai, X. D., Cao, Z. G., Wang, Y., Yu, Z. H., Zhang, X. F., and Li, C. N. (2013). Crop segmentation from images by morphology modeling in the CIE L*a*b* color space. *Computers and electronics in agriculture*.

Bastiaans, L., Paolini, R., and Baumann, D. (2008). Focus on ecological weed management: what is hindering adoption? *Weed Research*, 48(6):481–491.

Benaragama, D., Shirtliffe, S. J., Johnson, E. N., Duddu, H. S. N., and Syrovy, L. D. (2016). Does yield loss due to weed competition differ between organic and conventional cropping systems? *Weed Research*, 56(4):274–283.

Boland, G. J. and Hall, R. (1994). Index of plant hosts of sclerotinia sclerotiorum. *Canadian Journal of Plant Pathology*, 16(2):93–108.

Boström, U., Anderson, L. E., and Wallenhammar, A. C. (2012). Seed distance in relation to row distance: Effect on grain yield and weed biomass in organically grown winter wheat, spring wheat and spring oats. *Field Crops Research*, 134:144–152.

Brandsæter, L., Mangerud, K., Helgheim, M., Protection, T. B. C., and 2017, u. (2017). Control of perennial weeds in spring cereals through stubble cultivation and mouldboard ploughing during autumn or spring. *Elsevier*, 98:16–23.

Brandsæter, L. O., Mangerud, K., and Rasmussen, J. (2012). Interactions between pre- and post-emergence weed harrowing in spring cereals. *Weed Research*, 52(4):338–347.

Buddingh, C. and Buddingh, M. (1963). Rotary weeder. US Patent 3,082,829. Washington, DC: U.S. Patent and Trademark Office.

Bundesministerium für Ernährung und Landwirtschaft (BMEL) (2017). Rückstände von Pflanzenschutzmitteln - Gesundheit geht vor. Retrieved August 14, 2019. From https://www.bmel.de/SharedDocs/Bilder/Cover/Pflanzenschutzmittel-Rueckstaende.jpg.

Calicioglu, O., Flammini, A., Bracco, S., Bellù, L., and Sims, R. (2019). The future challenges of food and agriculture: An integrated analysis of trends and solutions. *Sustainability (Switzerland)*, 11(1).

Chambers, R. (1879). *Book of days*. J.B. Lippincott & Co, Philadelphia.

Champion, G. T., Froud-Williams, R. J., and HOLLAND, J. M. (1998). Interactions between wheat (Triticum aestivum L.) cultivar, row spacing and density and the effect on weed suppression and crop yield. *Annals of Applied Biology*, 133(3):443–453.

Chauvel, B., Guillemin, J. P., Gasquez, J., and Gauvrit, C. (2012). History of chemical weeding from 1944 to 2011 in France: Changes and evolution of herbicide molecules. *Crop Protection*, 42:320–326.

Christian, M. L. (1993). Survival of Wheat Streak Mosaic Virus in Grass Hosts in Kansas from Wheat Harvest to Fall Wheat Emergence. *Plant Disease*, 77(3):239.

CLAAS E-Systems (2019). Culti Cam. Retrieved September 6, 2019. From http://www.claas-e-systems.com/de/oem-produkte/culti-cam/.

Coble, H. D. and Schroeder, J. (2016). Call to Action on Herbicide Resistance Management. *Weed Science*, 64(sp1):661–666.

Cooper, J. and Dobson, H. (2007). The benefits of pesticides to mankind and the environment. *Crop Protection*, 26:1337–1348.

Dedousis, A. P., Godwin, R. J., O'Dogherty, M. J., Tillett, N. D., and Grundy, A. C. (2007). Inter and intra-row mechanical weed control with rotating discs. *Precision Agriculture*, 6:493–499.

Dierauer, H. and Stöppler-Zimmer, H (1994). *Unkrautregulierung ohne Chemie.* Ulmer, Stuttgart.

Diprose, M. F. and Benson, F. A. (1984). Electrical methods of killing plants.

Duke, S. O. (2012). Why have no new herbicide modes of action appeared in recent years? *Pest Management Science*, 68(4):505–512.

Duke, S. O. and Powles, S. B. (2008). Glyphosate: A once-in-a-century herbicide. *Pest Management Science*, 64(4):319–325.

E. Hahn (1910). *Niederer Ackerbau oder Hackbau?* Globus, 1 edition.

Einböck GmbH & CoKG (2017). UNIVERSAL ROW CROP CULTIVATION TECHNOLOGY, ROW-GUARD. Retrieved September 20, 2019. From https://www.einboeck.at/.

European Commission (2019). EU Pesticide Database. Retrieved Juli 19, 2019. From http://ec.europa.eu/food/plant/pesticides/eu-pesticides-database/public/.

Fahad, S., Hussain, S., Chauhan, B. S., Saud, S., Wu, C., Hassan, S., Tanveer, M., Jan, A., and Huang, J. (2015). Weed growth and crop yield loss in wheat as influenced by row spacing and weed emergence times. *Crop Protection*, 71:101–108.

Fehr, B. and Gerrish, J. (1995). Vision-guided row-crop follower. *Applied Engineering in agriculture*, 11:613–620.

Frieben, B. (1990). Bedeutung des organischen Landbaus für den Erhalt von Ackerwildkräutern. *Natur und Landschaft*, 65.7(8):379–382.

Garford Farm Machinery Ltd (2019). Robocrop. Retrieved September 20, 2019. From https://garford.com/wp-content/uploads/2018/07/Robocrop-Console.pdf.

Gerhards, R. and Christensen, S. (2003). Real-time weed detection, decision making and patch spraying in maize, sugarbeet, winter wheat and winter barley. *Weed Research*, 43(6):385–392.

Graf Littichau, G. and Liegnitz, B. (1941). Hoeing Machine.

Gressel, J. and Segel, L. A. (1978). The paucity of plants evolving genetic resistance to herbicides: Possible reasons and implications. *Journal of Theoretical Biology*, 75(3):349–371.

Griepentrog, H., Norremark, M., and Nielsen, J. (2006). Autonomous intra-row rotor weeding based on GPS. In *Proceedings of the CIGR World Congress Agricultural Engineering for a Better World*, volume 37, page 7, Bonn.

Grigg, D. B. (1975). The World's Agricultural Labour Force 1800-1970. *Geography*, 60(3):194–202.

Grovum, M. A. and Zoerb, G. C. (1970). An automatic guidance system for farm tractors. *Transactions of the ASAE*, 13(5):565–573.

Hague, T., Marchant, J. A., and Tillett, N. D. (2000). Ground based sensing systems for autonomous agricultural vehicles. *Computers and Electronics in Agriculture*, 25:11–28.

Hakansson, S. (1984). Row spacing, seed distribution in the row, amount of weedsinfluence on production in stands of cereals. In *25th Swedish Weeds Conference*, pages 17–34, Uppsala.

Hamill, A. S., Holt, J. S., and Mallory-Smith, C. A. (2004). Contributions of Weed Science to Weed Control and Management. *Weed Technology*, 18(1):1563–1565.

Hatcher, P. E. and Melander, B. (2003). Combining physical, cultural and biological methods: prospects for integrated non-chemical weed management strategies. *Weed Research*, 43(5):303–322.

Heap, I. (2019). The International Survey of Herbicide Resistant Weeds. Retrieved July 14, 2019. From http://www.weedscience.org/.

Heitefuss, R. K., Obst, A., and Reschke, M. (2000). *Pflanzenkrankheiten und Schädlinge im Ackerbau*. Verlags Union Agrar, 4 edition.

Hoban, T. J. (1998). Trends in consumer attitudes about agricultural biotechnology. *AgBioForum*, 1(1):3–7.

Holliday, R. (1963). The effect of row width on the yield of cereals. *Field Crop Abstr.*, 16:71–81.

Hough, P. V. C. (1962). Method and means for recognizing complex patterns. U.S. Patent No. 3,069,654. Washington, DC: U.S. Patent and Trademark Office.

International Labour Organization (2018). ILOSTAT database. Retrieve June 29, 2019. From https://ilostat.ilo.org/data/.

Jabran, K. and Chauhan, B. S. (2018). *Non-chemical weed control*. Academic Press, Elsevier, London, 1 edition.

Jacobsen, S. E., Christiansen, J. L., and Rasmussen, J. (2010). Weed harrowing and inter-row hoeing in organic grown quinoa (Chenopodium quinoa Willd.). *Outlook on Agriculture*, 39(3):223–227.

Jahns, G. (1976). Automatic Steering of Farm Vehicles. *Agric. Eng. Dept. Series*, 1:3–16.

Jahns, G. (1983). Automatic guidance in agriculture: A review. *ASAE paper NCR*, pages 83–404.

Jensen, R. K., Rasmussen, J., and Melander, B. (2004). Selectivity of weed harrowing in lupin. *Weed Research*, 44(4):245–253.

Jia, J., Krutz, G., and Gibson, H. (1990). Corn plant locating by image processing. *Optics in Agriculture*, 1379:246–253.

Johnson, E. (2002). Selectivity - An Important Concept in Mechanical Weed Control. *Research Report 2002, Canada-Saskatchewan Agri-Food Innovation Fund*.

Juroszek, P. and Gerhards, R. (2004). Photocontrol of weeds. *Journal of Agronomy and Crop Science*, 190(6):402–415.

Kalman, R. (1960). A New Approach to Linear Filtering and Prediction Problems. *Transactions of the ASME–Journal of Basic Engineering*, 82:35–45.

Kawasaki, K., Itokawa, N., and Ito, S. (1981). Studies on driverless transport vehicle guided by drain gutters on a steep slope citrus orchard, 2: Automatic steering performance for different running devices. *Bulletin of the Shikoku Agricultural Experiment Station (Japan)*, pages 143–176.

Keicher, R. and Seufert, H. (2000). Automatic guidance for agricultural vehicles in Europe. *Computers and Electronics in Agriculture*, 25(1-2):169–194.

Kirchhoff, C. and Duelks, A. (2019). Soil-working device; method for the height guidance of at least one finger-weeder tool above a row in a row plantation. US Patent 10,238,021. Washington, DC: U.S. Patent and Trademark Office.

Kolpin, D. W., Thurman, E. M., and Linhart, S. (1998). The environmental occurrence of herbicides: the importance of degradates in ground water. *Archives of Environmental Contamination and Toxicology*, 35(3):385–390.

Korres, N. E. and Froud-Williams, R. J. (2002). Effects of winter wheat cultivars and seed rate on the biological characteristics of naturally occurring weed flora. *Weed Research*, 42(6):417–428.

Kouwenhoven, J. K. (1997). Intra-row mechanical weed control - Possibilities and problems. *Soil and Tillage Research*, 41(1-2):87–104.

Kunz, C., Weber, J. F., and Gerhards, R. (2015). Benefits of Precision Farming Technologies for Mechanical Weed Control in Soybean and Sugar Beet—Comparison of Precision Hoeing with Conventional Mechanical Weed Control. *Agronomy*, 5(2):130–142.

Kunz, C., Weber, J. F., and Gerhards, R. (2016). Comparison of different mechanical weed control strategies in sugar beets. *27. Deutsche Arbeitsbesprechung über Fragen der Unkrautbiologie und -bekämpfung*, 1962:446–451.

Kurstjens, D. A. G. and Kropff, M. J. (2001). The impact of uprooting and soil-covering on the effectiveness of weed harrowing. *Weed Research*, 41(3):211–228.

Lemerle, D., Verbeek, B., Cousens, R. D., and Coombes, N. E. (1996). The potential for selecting wheat varieties strongly competitive against weeds. *Weed Research*, 36(6):505–513.

Lötjönen, T. and Mikkola, H. (2000). Three mechanical weed control techniques in spring cereals. *Agricultural and Food Science*, 9:269–278.

Machinefabriek Steketee B.V. (2017). Steketee IC-Light. Retrieved September 21, 2019. From https://www.steketee.com/de/steketee-ic-light/.

Marchant, J. A. and Brivot, R. (1995). Real-Time Tracking of Plant Rows Using a Hough Transform. *Real-Time Imaging*, 1(5):363–371.

Martindale, W. and Trewavas, A. (2008). Fuelling the 9 billion. *Nature Biotechnology*, 26(10):1068–1070.

Maschienenfabrik Schmotzer GmbH (2018). Kamerasteuerung Schmotzer Okio. Retrieved September 6, 2019. From https://schmotzer.de/files/xsvzc.pdf.

Massa, D., Kaiser, Y., Andújar-Sánchez, D., Carmona-Alférez, R., Mehrtens, J., and Gerhards, R. (2013). Development of a Geo-Referenced Database for Weed Mapping and Analysis of Agronomic Factors Affecting Herbicide Resistance in Apera spica-venti L. Beauv. (Silky Windgrass). *Agronomy*, 3(1):13–27.

Mathiassen, S. K., Bak, T., Christensen, S., and Kudsk, P. P. (2006). The Effect of Laser Treatment as a Weed Control Method. *Biosystems Engineering*, 95(4):497–505.

Melander, B. (2006). Current achievements and future directions of physical weed control in Europe. In *Third International Conference on Non-chemical Crop Protection Methods*, pages 49–58, Lille.

Melander, B., Cirujeda, A., and Jørgensen, M. H. (2003). Effects of inter-row hoeing and fertilizer placement on weed growth and yield of winter wheat. *Weed Research*, 43(6):428–438.

Melander, B., Holst, N., Rasmussen, I. A., and Hansen, K. (2012). Direct control of perennial weeds between crops-Implications for organic farming. *Crop Protection*, 40:36–42.

Melander, B., Rasmussen, I. A., and Bàrberi, P. (2005). Integrating physical and cultural methods of weed control— examples from European research. *Weed Science*, 53(3):369–381.

Merfield, C. N. (2019). Integrated Weed Management in Organic Farming. In *Organic Farming*, pages 117–180. Elsevier.

Millet, M., Bertrand, F., and Bedos, C. (2016). Atmospheric concentrations and volatilisation fluxes of two herbicides applies on maize. *Fresenius Environmental Bulletin*, 12(7):675–679.

Monaco, T., Weller, S., and Ashton, F. (2002). *Weed science: principles and practices*. John Wiley & Sons, New York, 4 edition.

Mousazadeh, H. (2013). A technical review on navigation systems of agricultural autonomous off-road vehicles. *Journal of Terramechanics*, 50(3):211–232.

Moyer, J., Roman, E., Lindwall, C., and Blackshaw, R. (1994). Weed management in conservation tillage systems for wheat production in north and south america. *Crop Protection*, 13(4):243–259.

Mülle, G. and Heege, H. J. (1981). Kornverteilung über die Fläche und Ertrag bei Getreide. *Z. Acker-Pflanzenb*, 150:97–112.

Murchie, E. H., Pinto, M., and Horton, P. (2009). Agriculture and the new challenges for photosynthesis research. *New Phytologist*, 181(3):532–552.

Neumann, M., Schulz, R., Schäfer, K., and Müller, W. (2002). The significance of entry routes as point and non-point sources of pesticides in small streams. *Water Research*, 36(4):835–842.

Nitschke, L. and Schüssler, W. (1998). Surface water pollution by herbicides from effluents of waste water treatment plants. *Chemosphere*, 36(1):35–41.

Nørremark, M., Griepentrog, H. W., Nielsen, J., and Søgaard, H. T. (2012). Evaluation of an autonomous GPS-based system for intra-row weed control by assessing the tilled area. *Precision Agriculture*, 13(2):149–162.

Oerke, E. C. (2006). Crop losses to pests. *The Journal of Agricultural Science*, 144(01):31.

Onyango, C. M. and Marchant, J. (2003). Segmentation of row crop plants from weeds using colour and morphology. *Computers and Electronics in Agriculture*, 39(3):141–155.

Orr, J. R. and Radley, W. J. (1913). Automatic weed-cutter and hoeing attachment for cultivators and plows. US Patent 1,050,993. Washington, DC: U.S. Patent and Trademark Office.

Panizzi, A. R. (1997). Wild Host of Penetatomids: Ecological Significance and Role in Their Pest Status on Crops. *Annual Review of Entomology*, 42(1):99–122.

Phillips, L. W. (1949). Automatic steering attachment for tractors. US Patent 2,465,660. Washington, DC: U.S. Patent and Trademark Office.

Pietsch, J., Schmidt, W., Sacher, F., Fichtner, S., and Brauch, H. J. (1995). Pesticides and other organic micro pollutants in the river Elbe. *Fresenius' Journal of Analytical Chemistry*, 353(1):75–82.

Poulsen, F. (2016). FIELD VISION SYSTEMS. Retrieved September 1, 2019. From http://www.visionweeding.com/field-vision-system/.

Poulsen, F. (2018). The ROBOVATOR. Retrieved September 4, 2019. From http://www.visionweeding.com/robovator/.

Prince, J. M., Shaw, D. R., Givens, W. A., Newman, M. E., Owen, M. D. K., Weller, S. C., Young, B. G., Wilson, R. G., and Jordan, D. L. (2012). Benchmark Study: III. Survey on Changing Herbicide Use Patterns in Glyphosate-Resistant Cropping Systems. *Weed Technology*, 26:536–542.

Pullen, D. W. and Cowell, P. A. (1997). An evaluation of the performance of mechanical weeding mechanisms for use in high speed inter-row weeding of arable crops. *Journal of Agricultural and Engineering Research*, 67(1):27–34.

Pullen, D. W. and Cowell, P. A. (2000). Prediction and experimental verification of the hoe path of a rear-mounted inter-row weeder. *Journal of Agricultural and Engineering Research*, 77(2):137–153.

Pullen, D. W. and Cowell, P. A. (2006). The Effect of Implement Geometry on the Hoe Path of a Steered Rear-mounted Inter-row Weeder. *Biosystems Engineering*, 94(3):373–386.

Rasmussen, I. A. (2004). The effect of sowing date, stale seedbed, row width and mechanical weed control on weeds and yields of organic winter wheat. *Weed Research*, 44(1):12–20.

Rasmussen, J. (1991). A model for prediction of yield response in weed harrowing. *Weed Research*, 31(February):401–408.

Rasmussen, J. (1992). Testing harrows for mechanical control of annual weeds in agricultural crops. *Weed Research*, 32(4):267–274.

Rasmussen, J. (1997). Hvilken langfingerharve er bedst til korn? In *Pesticider Og Miljø, Ukrudt*, pages 203–213. Danmarks Jordbrugsforskning, Forskningscenter Foulum.

Rasmussen, J., Bibby, B. M., and Schou, A. P. (2008). Investigating the selectivity of weed harrowing with new methods. *Weed Research*, 48(6):523–532.

Rasmussen, J., Nørremark, M., and Bibby, B. M. (2007). Assessment of leaf cover and crop soil cover in weed harrowing research using digital images. *Weed Research*, 47(4):299–310.

Rasmussen, J. and Svenningsen, T. (1995). Selective weed harrowing in cereals. *Biological Agriculture and Horticulture*, 12(1):29–46.

Reid, J. and Searcy, S. (1986). Detecting crop rows using the hough transform. *American Society of Agricultural Engineers. Microfiche collection (USA)*.

Rueda-Ayala, V., Peteinatos, G., Gerhards, R., and Andújar, D. (2015). A non-chemical system for onlineweed control. *Sensors (Switzerland)*, 15(4):7691–7707.

Rueda-Ayala, V. P., Rasmussen, J., Gerhards, R., and Fournaise, N. E. (2011). The influence of post-emergence weed harrowing on selectivity, crop recovery and crop yield in different growth stages of winter wheat. *Weed Research*, 51(5):478–488.

Ruiz-Ruiz, G., Gómez-Gil, J., and Navas-Gracia, L. M. (2009). Testing different color spaces based on hue for the environmentally adaptive segmentation algorithm (EASA). *Computers and Electronics in Agriculture*, 68:88–96.

Rydberg, N. T. and Milberg, P. (2000). A Survey of Weeds in Organic Farming in Sweden. *Biological Agriculture & Horticulture*, 18(2):175–185.

Rydberg, T. (1994). Weed harrowing—the influence of driving speed and driving direction on degree of soil covering and the growth of weed and crop plants. *Biological Agriculture and Horticulture*, 10(3):197–205.

Salzman, F. P. and Renner, K. A. (1992). Response of Soybean to Combinations of Clomazone, Metribuzin, Linuron, Alachlor, and Atrazine. *Weed Technology*, 6(4):922–929.

Shaw, D. R. (2016). The "Wicked" Nature of the Herbicide Resistance Problem. *Weed Science*, 64(sp1):552–558.

Slaughter, D. C., Chen, P., and Curley, R. G. (1999). Vision Guided Precision Cultivation. *Precision Agriculture Precision Agriculture*, 1(2):199–217.

Slaughter, D. C., Giles, D. K., and Downey, D. (2008). Autonomous robotic weed control systems: A review. *Computers and Electronics in Agriculture*, 61(1):63–78.

Squillace, P. J. and Thurman, E. M. (1992). Herbicide Transport in Rivers: Importance of Hydrology and Geochemistry in Nonpoint-Source Contamination. *Environmental Science and Technology*, 26(3):538–545.

Stanhope, T., Viacheslav, I., and Roux, J. (2014). Computer vision guidance of field cultivation for organic row crop production. In *2014 ASABE Annual International Meeting*, pages 1–9, Montreal.

Switzer, C. (1957). The existence of 2,4-D-resistant strains of wild carrot. In *Proceedings of the Eleventh Northeastern Weed Control Conference*, pages 315–318, New York.

Terpstra, R. and Kouwenhoven, J. (1981). Inter-row and intra-row weed control with a hoe-ridger. *Journal of Agricultural Engineering Research*, 26(2):127–134.

Tian, L. F. and Slaughter, D. C. (1998). Environmentally adaptive segmentation algorithm for outdoor image segmentation. *Computers and Electronics in Agriculture*, 21(3):153–168.

Tillett, N. (2005). Cost effective weed control in cereals using vision guided inter-row hoeing and band sparying systems. Technical Report 370, Tillett and Hague Technology Limited, Greenfield, Bedfordshire.

Tillett, N. D. (1991). Automatic guidance sensors for agricultural field machines:A review. *Journal of Agricultural Engineering Research*, 50(C):167–187.

Tillett, N. D. and Hague, T. (1999). Computer-vision-based hoe guidance for cereals - An initial trial. *Journal of Agricultural and Engineering Research*, 74(3):225–236.

Tillett, N. D., Hague, T., Grundy, A. C., and Dedousis, A. P. (2008). Mechanical within-row weed control for transplanted crops using computer vision. *Biosystems Engineering*, 99(2):171–178.

Tillett, N. D., Hague, T., and Miles, S. J. (2002). Inter-row vision guidance for mechanical weed control in sugar beet. *Computers and Electronics in Agriculture*, 33(3):163–177.

Tillett and Hague Technology Ltd. (2018). Inter-Row Vision Guidance. Retrieve August 12, 2019. From http://www.thtechnology.co.uk/products.html.

Tørresen, K. S., Skuterud, R., Tandsæther, H. J., and Hagemo, M. B. (2003). Long-term experiments with reduced tillage in spring cereals. I. Effects on weed flora, weed seedbank and grain yield. *Crop Protection*, 22(1):185–200.

Van der Schans, D., Bleeker, P., Molendijk, L., Plentinger, M., Van Der Weide, R., Lotz, L., Bauermeister, R., Total, R., and Baumann, D. T. (2006). *Practical weed control in arable farming and outdoor vegetable cultivation without chemicals*, volume 352. Wageningen UR, Applied Plant Research.

Van der Weide, R. and Bouma, E. (1997). Vorstgevaar valt mee: Mechanisch onkruid bestrijden niet per se riskant. *Boerderij/akkerbouw*, 8.

Van Der Weide, R. Y., Bleeker, P. O., Achten, V. T. J. M., Lotz, L. A. P., Fogelberg, F., and Melander, B. (2008). Innovation in mechanical weed control in crop rows. *Weed Research*, 48(3):215–224.

Van Oost, K., Govers, G., de Alba, S., and Quine, T. A. (2006). Tillage erosion: A review of controlling factors and implications for soil quality. *Progress in Physical Geography*, 30(4):443–466.

Van Overbeek, J. (1947). Use of synthetic hormones as weed killers in tropical agriculture. *Economic Botany*, 1(4):446–459.

Waibel, H., Fleischer, G., Becker, H., and Runge-Metzger, A. (1998). Kosten und Nutzen des chemischen Pflanzenschutzes in der deutschen Landwirtschaft aus gesamtwirtschaftlicher Sicht. In Becker, H. and Runge-Metzger, A., editors, *Agrarökonomische Monographien und Sammelwerke*, pages 74–76. Wiss.-Verlag Vauk, Kiel.

Walker, P. (1983). Crop losses: The need to quantify the effects of pests, diseases and weeds on agricultural production. *Agriculture, Ecosystems & Environment*, 9(2):119–158.

Walker, R. H. (1995). Preventive Weed Management. In Smith, E., editor, *Handbook of Weed Management Systems*, chapter 3, page 35. Marcel Decker, Inc., New York, 2 edition.

Wegler, R. (2013). *Chemie der Pflanzenschutz-und Schädlingsbekämpfungsmittel: Band 2: Fungizide· Herbizide· Natürliche· Pflanzenwuchsstoffe· Rückstandsprobleme.* Springer-Verlag, Berlin, 5 edition.

Welsh, J. P., Bulson, H. A. J., Stopes, C. E., Froud-Williams, R. J., and Murdoch, A. J. (1999). The critical weed-free period in organically-grown winter wheat. Technical report, Department of Agricultural Botany, The University of Reading.

Wilson, B. J. and Wright, K. J. (1990). Predicting the growth and competitive effects of annual weeds in wheat. *Weed Research*, 30(3):201–211.

Wilson, J. (2000). Guidance of agricultural vehicles — a historical perspective. *Computers and Electronics in Agriculture*, 25(1-2):3–9.

Wise, J. and Whalon, M. (2009). A Systems Approach to IPM Integration, Ecological Assessment and Resistance Management in Tree Fruit Orchards. In *Biorational Control of Arthropod Pests*, pages 325–345. Springer Netherlands, Dordrecht.

Woebbecke, D., Meyer, G., and Von Bargen, K. (1995). Color indices for weed identification under various soil, residue, and lighting conditions. *Transactions of the ASAE*, 38(1):259–269.

Young, F. L., Ogg Jnr, A. G., Papendick, R. I., Thill, D. C., and Alldredge, J. R. (1994). Tillage and weed management affects winter wheat yield in an integrated pest management system. *Agronomy Journal*, 86(1):147–154.

Zhang, X. and Chen, Y. (2017). Soil disturbance and cutting forces of four different sweeps for mechanical weeding. *Soil and Tillage Research*, 168:167–175.

Zimdahl, R. L. (2007). *Fundamentals of Weed Science*. Academic Press, Elsevier, London, 5 edition.

Zoschke, A. and Quadranti, M. (2002). Integrated weed management: Quo vadis? *Weed Biology and Management*, 2(1):1–10.

Appendix A

Declaration in lieu of an oath on independent work

1. The dissertation submitted on the topic *Development of Hoeing in Narrow Seeded Cereals with a Camera Row Guidance* is work done independently by me.

2. I only used the sources and aids listed and did not make use of any impermissible assistance from third parties. In particular, I marked all content taken word-for-word or paraphrased from other works.

3. I did not use the assistance of a commercial doctoral placement or advising agency.

4. I am aware of the importance of the declaration in lieu of oath and the criminal consequences of false or incomplete declarations in lieu of oath.

5. I confirm that the declaration above is correct. I declare in lieu of oath that I have declared only the truth to the best of my knowledge and have not omitted anything.

Stuttgart, 23.01.2020	Benjamin Kollenda
Place, Date	Signature

Appendix B

Curriculum vitae

Personal Details

Benjamin Kollenda
Gorch-Fock-Str. 25
70619 Stuttgart
Mobil: +49-1577-9796670
E-mail: benjamin.kollenda@gmx.de
Born the 5th of August 1988
in Filderstadt, Germany

Education and Training

2017-2020	Doctoral Candidate: University of Hohenheim
	Faculty of Agricultural Science
	Doctoral thesis:
	Development of Hoeing in Narrow Seeded Cereals with a Camera Row-Guidance
2014-2016	Master of Science: Plant Production Systems
	University of Hohenheim
	Department of Weed Science
	Master thesis:
	Further Development of a Logarithmic Sprayer for Dose-Response Experiments

---

2010-2014	Bachelor of Science: Agricultural Biology
	University of Hohenheim
	Department of Phytopathology
	Bachelor thesis:
	Der Effektorkandidat PpEC23:
	Überexpression und Pulldown

Career history

2016	Internship at Monsanto Australia Ltd.
2010-2015	Part-time employee at Johanniter-Unfall-Hilfe e.V., Stuttgart
2009-2010	Employee at Johanniter-Unfall-Hilfe e.V., Stuttgart
2008-2009	Civil service as Paramedic at Johanniter-Unfall-Hilfe e.V., Stuttgart

Languages

German	Mother tongue
English	Fluent
French	Basic